Code of Practice for Project Management for the Built Environment

Code of Practice for Project Management for the Built Environment

Sixth Edition

THE CHARTERED INSTITUTE OF BUILDING

This sixth edition first published 2022
© 2022 John Wiley & Sons Ltd

Edition History
Blackwell Science Ltd 1e, 1992; 2e, 1996
Blackwell Publishing Ltd 3e, 2002
John Wiley & Sons Ltd 4e, 2010; 5e, 2014
all as Code of Practice for Project Management for Construction and Development

Registered Offices
John Wiley & Sons, Inc., 111 River Street, Hoboken, NJ 07030, USA
John Wiley & Sons Ltd, The Atrium, Southern Gate, Chichester, West Sussex, PO19 8SQ, UK

Editorial Office
9600 Garsington Road, Oxford, OX4 2DQ, UK

For details of our global editorial offices, customer services, and more information about Wiley products visit us at www.wiley.com.

Wiley also publishes its books in a variety of electronic formats and by print-on-demand. Some content that appears in standard print versions of this book may not be available in other formats.

Limit of Liability/Disclaimer of Warranty
In view of ongoing research, equipment modifications, changes in governmental regulations, and the constant flow of information relating to the use of experimental reagents, equipment, and devices, the reader is urged to review and evaluate the information provided in the package insert or instructions for each chemical, piece of equipment, reagent, or device for, among other things, any changes in the instructions or indication of usage and for added warnings and precautions. While the publisher and authors have used their best efforts in preparing this work, they make no representations or warranties with respect to the accuracy or completeness of the contents of this work and specifically disclaim all warranties, including without limitation any implied warranties of merchantability or fitness for a particular purpose. No warranty may be created or extended by sales representatives, written sales materials or promotional statements for this work. The fact that an organization, website, or product is referred to in this work as a citation and/or potential source of further information does not mean that the publisher and authors endorse the information or services the organization, website, or product may provide or recommendations it may make. This work is sold with the understanding that the publisher is not engaged in rendering professional services. The advice and strategies contained herein may not be suitable for your situation. You should consult with a specialist where appropriate. Further, readers should be aware that websites listed in this work may have changed or disappeared between when this work was written and when it is read. Neither the publisher nor authors shall be liable for any loss of profit or any other commercial damages, including but not limited to special, incidental, consequential, or other damages.

Library of Congress Cataloging-in-Publication Data

Names: Chartered Institute of Building (Great Britain), author.
Title: Code of practice for project management for the built environment /
  CIOB.
Other titles: Code of practice for project management for construction and
  development
Description: Sixth edition. | Hoboken, NJ, USA: Wiley-Blackwell, 2022. |
  Revised edition of: Code of practice for project management for
  construction and development. | Includes bibliographical references and
  index.
Identifiers: LCCN 2022013255 (print) | LCCN 2022013256 (ebook) | ISBN
  9781119715139 (paperback) | ISBN 9781119715153 (adobe pdf) | ISBN
  9781119715214 (E-pub)
Subjects: LCSH: Building–Superintendence. | Project management.
Classification: LCC TH438 .C626 2022 (print) | LCC TH438 (ebook) | DDC
  690.068–dc23/eng/20220407
LC record available at https://lccn.loc.gov/2022013255
LC ebook record available at https://lccn.loc.gov/2022013256

Cover design: Wiley
Cover image: © Sergey Nivens/Shutterstock

Set in 10.5/13.5pt Franklin Gothic by Straive, Pondicherry, India

SKYCF8CA532-0F21-404F-9AE5-1446B54BB999_051322

Contents

Contents

Contents

Foreword

The fifth edition of the Code of Practice was published in 2014 and there have been many challenges and changes within the industry since then, which had to be reflected in the sixth edition.

When we set out on the task of developing the sixth edition, we initially thought that we could carry out an update of the fifth edition. We very soon realised that there has been so much change in the industry between 2014 and 2022 that a significant rewrite and restructure of the Code of Practice would be required. In fact, I think it's fair to say that in the last ten years we have seen change – the speed of which is unprecedented, thus the need for a restructuring and rewrite of the Code rather than an update of what we had in the fifth edition.

The overarching main theme for my Presidential year is the role of the client, and I have a firm belief that clients should be the force for positive change. For the change in the industry to be maintained in a sustainable manner, the role of the Client Project Manager must not be under-estimated.

I have the ambition for CIOB to be the professional body of choice for clients globally, and I believe that this sixth edition will be of great assistance to all Project Managers representing clients of all sizes and organisational types.

Equally the sixth edition serves as the 'go-to' Code of Practice for all Project Managers regardless of their backgrounds and non-client role within the Built Environment processes.

The importance of good project management has never been so vital. The industry has many challenges as it strives to change its image and reputation, increase productivity, improve quality and building safety in addition to a modernisation programme, which will bring more off-site manufacturing alongside the role of digitisation. Sustainability is at the heart of the CIOB's corporate plan and as such, will be producing a separate document to complement the *Code of Practice for Project Management*, sixth edition, that specialises in climate and environmental risk matters.

I pay tribute to my colleagues at CIOB and others who have offered advice and guidance as the sixth edition was developed. My special thanks and appreciation go to the working group for their huge efforts and diligence in commenting and contributing to this sixth edition. With their help, I hope that this version provides a positive contribution to all the challenges referred to above as the industry continues its journey of improvement in the complex world it operates in.

Mike Foy OBE MBA FCIOB FCMI
CIOB President 2021/2022

Acknowledgements

I'm delighted to present the latest edition of CIOB's Code of Practice for Project Management for the Built Environment.

Since the ground-breaking first edition of this Code of Practice was published in 1992, it has evolved over the years to meet the needs of built environment professionals in an ever-changing industry. This edition, the sixth, has been significantly revamped to reflect those changes and remain at the forefront of professional practice in construction project management.

This document is the result of a great deal of hard work from some talented and experienced people and I would like to thank them for their contribution. Particular acknowledgement and thanks must go to the working group, expertly chaired by Mike Foy OBE FCIOB (CIOB President 2021/2022) – I thank them for their expertise and commitment and for producing this excellent publication. My gratitude also extends to my CIOB colleagues who contributed to the creation of this edition of the Code of Practice.

Finally, I would like to thank all our members and readers who take the time to stay at the forefront of their profession and lead by example with the adoption of best practice wherever possible.

I hope you enjoy the wealth of knowledge and practical guidance contained in our updated Code of Practice for Project Management for the Built Environment.

Caroline Gumble
Chief Executive

Working group for the revision of the *Code of Practice for Project Management* – Sixth Edition

Michael Foy OBE MBA FCIOB FCMI	Working Party Chair, HM Government – Department for Education
Christine Gausden RD FCIOB	Academic Delivery Manager – University College of Estate Management
Gildas André MBA MSc BSc (Hons) MAPM MCIOB	Director – GAN Advisory Services
David Churcher MBE	Director – Hitherwood Consulting
David Haimes MCIOB	Director Regional Investment Programme North – Highways England
Stan Hardwick FCIOB MIHEEM EurBE	Head of Clinical Contracts Management – Specsavers/Newmedica
Joanna Harris FRICS	UK&I Hard FM Ambassador, Sodexo
Dr Paul Sayer	Publisher – Wiley-Blackwell, John Wiley & Sons Ltd, Oxford
Dr Gina Al-Talal BEng, MPhil PhD, FCIOB	Code of Practice Project Lead (Technical), Head of Technical & Standards Development – CIOB
Adrian Montague BSc MA	Code of Practice Deputy Lead, Associate Director - Academy – CIOB
Tracey Clarke MSc Econ	Code of Practice Project Manager, Technical Assets Manager – CIOB

Code of Practice for Project Management – Technical Authors

Dr Ruth Murray-Webster HonFAPM	Director – Potentiality UK
John C A Hayes FRICS FCIOB MCIArb	Owner and Director – GMS Ltd

The following also contributed in development of the sixth edition of the *Code of Practice for Project Management*

Stephen Coppin BEng(Hons) MSc FCIOB CFIOSH FIIRSM FaPS CWIFM MInstRE	Strategic Technical Advisor – SJC Risk Management Solutions Ltd
Cesare McArdle LLB(Hons)	Legal Director – Herrington Carmichael LLP

Max Muncaster MSc MCIOB — Senior Lecturer in Building Services – Oxford Brookes University

Paul Nash MSc FCIOB — Past President, Chartered Institute of Building

Asmau Nasir MSc, MCIOB, MRICS — Director – Qunekt Ltd

David Stockdale MBA FRICS FIoD Cenv — Director, Briga Consulting Limited | A CIOB Chartered Building Consultancy

Terri Warren LLB(Hons) LLM(Hons) — Solicitor – Herrington Carmichael LLP

Dr Mike Webster BEng, MSc, PhD, DIC, CEng, FICE, FIStructE — Director – MPW R&R Ltd

Additional thanks

Andrew Boyle FCIOB — Project Director at Muse Development

Evangelos Maltezos CEng MICE CMILT MBA — Senior Project Manager at VINCI Construction UK (Taylor Woodrow)

Steven Thompson BSc MBA FRICS — Professional Groups Associate Director of the Built Environment RICS

Dr Shu-Ling Lu PhD FCIOB FHEA — Associate Professor of Construction and Project Management at University of Reading

Neil Lock MCIOB — Operations Director at Morgan Sindall

Scott Walkinshaw — Head of Knowledge at APM

Professor Lloyd Scott — Former Head of Knowledge in Sustainable Construction Practices research group at Technological University Dublin

List of figures and diagrams

List of tables

Introduction

Purpose

The purpose of this sixth edition of the *Code of Practice for Project Management for the Built Environment* is to bring together:

- best practice associated with the management of projects in the built environment, with
- the strategic imperatives of the sector to drive a 'step change' in performance in terms of health, safety, well-being, sustainability, quality, productivity and value and
- to assist users with guidance and advice related to efficient and effective project management.

The construction industry is advancing and accelerating as we incorporate improvements into our built environment focusing on the whole life cycle of an asset, including the implications of climate change. It is therefore imperative to continually monitor and validate key changes in UK Government policies and corporate strategies and that the CIOB *Code of Practice* is used to improve our construction operational capability.

The vision of the Chartered Institute of Building (CIOB) is to lead and inspire excellence in the built environment, focused on the CIOB Royal Charter. It is important to follow the national and internationally agreed standards, i.e. BREEAM, LEED, GreenStar etc. This will help reduce consumer demand for heavily polluting goods and services, the aim being to promote cleaner energy and transport systems with non-fossil fuels producing at least 60% of the required energy output by 2050 in order to achieve the required drastic reductions of carbon emissions. It could also be suggested that user/client/organisational behaviour/expectations need to change significantly to achieve this. This edition of the *Code of Practice* consolidates latest thinking on the part project management plays in achieving the vision, considering the whole life of assets. It provides a guide for practitioners at all stages of their professional development and career.

Core concepts

There are many different definitions in existence for the core concepts and ideas used throughout this *Code of Practice*, including what is meant by the

Code of Practice for Project Management for the Built Environment, Sixth Edition. Chartered Institute of Building.
© 2022 John Wiley & Sons Ltd. Published 2022 by John Wiley & Sons Ltd.

'built environment', what is a 'project' and what is meant by the term 'life cycle'. A description of core concepts is provided as follows:

Built environment

The term 'built environment' relates to man-made assets and infrastructure, regardless of client type, funding, size, scale or complexity. Built assets and infrastructure exist in transportation (road, rail, airports, maritime ports), power and utilities (nuclear, oil and gas, tidal lagoons, offshore wind, solar, water, electricity, telecommunications), natural defences (flood defences, dams), as well as buildings (homes, hospitals, schools, factories, warehouses, offices, hotels) and the parks, plazas and other spaces that create the environment in which people interact. All sectors continue to advance at a pace, a full and separate awareness of progress and improvements is imperative for an efficient and effective project manager.

Project

The term 'project' describes the multiple ways over the entire life cycle of a built environment asset, in which clients organise the work to create, repurpose and eventually retire built assets in order to achieve objectives and realise the desired value for end users, clients and funding bodies. Projects are delivered by temporary organisations: teams of people, from multiple organisations, working collaboratively in a structured way:

- to achieve defined objectives and quality standards

- in a context of competing time and cost constraints

- navigating significant uncertainty and risk

- to operate in an environmentally friendly manner

- to provide feedback and learn lessons

- to evaluate the performance of the project stakeholders

Life cycle

The term 'life cycle' refers to the key stages of the whole life of a built asset and the objectives and decisions associated with each stage. The *Code of Practice* is structured around eight life cycle stages that address the work necessary to identify, assess, define, design, implement, validate, operate and retire assets. The scope of some projects will address only parts of the asset life cycle, but the client nevertheless has the responsibility to decide how to organise the work across the whole life cycle, including the governance required to decide to move from one life cycle stage to the next. The life cycle is best considered as a closed loop as shown in Figure 0.1 with a final decision-point culminating at the end of life of an asset whether to (1) extend life by continuing to use and maintain to the original design or to (2) retire the original design and repurpose all or parts of the asset. The latter scenario would trigger a new project. It is important to map out what stages of the life cycle apply to your projects in the identification phase of your scheme. If relevant actions and project data are applied to each stage and continually monitored via updated programmes and plans, it will be possible to manage each element of the life cycle before moving onto the next stage. It is recommended that a validated and logic linked plan with a critical path is developed as early as possible in the project life cycle.

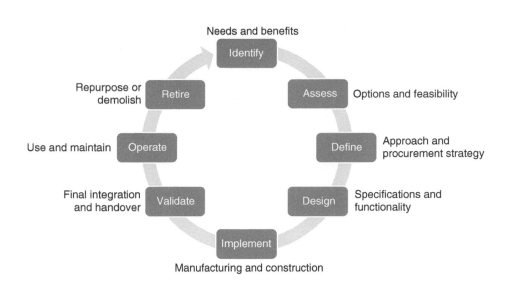

Figure 0.1 Project life cycle.

Different parts of the sector use different terminology for the life cycle stages as shown in Table 0.1. Some focus on the whole life of the asset, others on a more limited set of stages reflecting a particular technical discipline or specialist viewpoint. The rationale for the life cycle stages chosen is to reflect the broadest application of project management in the built environment and to focus the stage titles on the work undertaken in each stage.

It is important in the current climate to include an awareness of the sustainability life cycle. The below Table 0.2 highlights the stages and activities that should be considered in addition to the project life cycle stages described in the table above.

Sustainability overview Sustainability and sustainable development have advanced greatly since the fifth edition of the *Code of Practice*. Climate science has improved the understanding of how the planet has reacted negatively to industrial practices and unsustainable consumerism.

Sustainability encompasses the three core elements, namely economic, social and environmental. It is relevant that a project manager in the built environment appreciates these issues, familiarises themselves with innovative and more environmentally friendly ways of working and, where relevant, recommends or appoints the necessary expertise to ensure that the correct sustainability criteria and governance is embedded in projects and programmes of work.

The CIOB will be producing a separate document to complement the sixth edition of the *Code of Practice for Project Management* that specialises in climate and environmental risk matters.

As a minimum, the sustainability expert should be able to inform and advise from a sustainability perspective on the following criteria:

- Understanding the project's core and supporting processes (and sub-processes)
- Understanding and communicating the benefits of a sustainable supply chain
- Identifying the primary and secondary supply chain stakeholders

Table 0.1 Comparison of project life cycle stages

Sixth Edition of CIOB Code of Practice for Project Management for the Built Environment	Royal Institute of British Architects (RIBA) Plan of Work 2020	ISO 55000:2014 Asset management – Overview, principles and terminology	BS 6079:2019 Project management. Principles and guidance for the management of projects	Fifth Edition of CIOB Code of Practice for Project Management for the Built Environment (reference only)
1. **Identify:** needs and benefits. What are the needs and benefits for your specific project?	0. Strategic definition 1. Preparation and brief	Business Case	Investigation	Inception
2. **Assess:** options and feasibility (for the project(s))	2. Concept design			Feasibility
3. **Define:** approach and procurement strategy, logic linked to a defined programme	(programming and planning)		Development	Strategy
4. **Design:** specifications and functionality ensuring a coordinated design	3. Spatial coordination 4. Technical design	Create or acquire		Pre-construction
5. **Implement:** manufacture and construction applying robust quality assurance and quality control procedures	5. Manufacture and construction			Construction
6. **Validate:** integrate and handover with full administration procedures	6. Handover			Testing and commissioning
7. **Operate:** use and maintain to meet the clients' expectations	7. Use	Operate and maintain	Operation	Completion, handover and operation Post-completion review and in use
8. **Retire:** repurpose or demolish for a new function		Dispose or replace	Upgrade, or disposal/ retirement/withdrawal	

Table 0.2 Stages in the sustainability life cycle

Sixth Edition of the CIOB *Code of Practice for Project Management*	SOCIAL Sustainability activities	ECONOMIC Sustainability activities	ENVIRONMENTAL Sustainability activities
1. **Identify**: needs and benefits	Establish sustainability goals from international, national and local government strategies, from the client and end-user perspective or (via your own) business requirements	Establish sustainability goals from international, national and local government strategies, from the client and end-user perspective or (via your own) business requirements	Establish sustainability goals and establish your environmental mandate from international, national and local government strategies, from the client and end-user perspective or (via your own) business requirements
	Identify project impact on social needs from local investigations. This should include an analysis of your Corporate Social Responsibility (CSR) strategy	Identify project impact on the economy from local investigations	Identify project impact on the environment from local investigations
			Review (your) existing organisational (environmental) policy
2. **Assess**: options and feasibility	Carry out impact assessments Review legal requirements Establish best practice including consideration for stakeholders Consider type of users/stakeholders Consider financial incentives/funding options with stakeholder involvement		
3. **Develop**: approach and procurement strategy	Produce a sustainability plan to include sustainability in procurement criteria		
4. **Design**: specifications and functionality	Co-ordinate, review and update sustainability requirements		
5. **Implement**: manufacture and construct	Quality criteria, monitoring and benchmarking incorporated into a quality assurance and quality control process		
6. **Validate**: integrate and handover	Information and education including continual information and education flow		
7. **Operate**: use and maintain	Monitor all data management and review results for incorporation into operational and maintenance documentation		
8. **Retire**: repurpose or demolish	Review and implement findings as necessary		

- Undertaking a stakeholder analysis and establishing who has the greatest impact on the sustainable supply chain (using techniques such as Pareto analysis)

- Knowing where products are coming from and controlling product/component variety

- Benchmarking performance using science-based targets

- Treating a sustainable supply chain as a systemic risk to the project
- Creating a communication system to capture lessons learned and to facilitate the exchange of best practice
- Recommend and advise in the use of clean technology energy in the construction life cycle

Environmental mandates (including social value and carbon net zero aspirations)

Economic, social and environmental performance and impact is a critical client consideration. These elements of sustainability and sustainable development include sustainable financing, which takes into account the environmental and social performance of the asset. They may also include requirements for embodied and operational carbon emissions and energy consumption, as well as prescribing requirements for environmental impact on the local topography or adjacent area. Lastly, outcomes in terms of the local community, such as providing employment and training opportunities or the use of local supply chain, may be determined.

A sustainability mandate for the project will support the management framework for the planning and implementation of construction activities in accordance with the sustainability commitments of the organisation, the project context, funders, project end users or any other stakeholders.

The sustainability mandate will influence key design parameters relating to sustainability, performance and operational technologies. It should also outline the overall environmental management criteria including what are the key success factors (for the project(s)) in terms of sustainability management. It is the role of the project manager to debate, evolve, prepare and manage all required sustainability mandates, noting that the environment is only one element of sustainability. This will require project managers to understand and have knowledge of environmental science/energy management of built assets, climate change science or biodiversity sufficiently enough to have meaningful discussions with the project stakeholders at all life cycle stages.

Strategic drivers

Since publication of the fifth edition, strategic drivers and priorities have been influenced by a number of occurrences and reports across the world. In 2013, the UK Government issued the 2025 vision and strategy for construction.[1] The vision was to adopt 'smart construction', which is low carbon and sustainable, supported by digital design to improve performance.

Management of the impact of organisational activities on the natural environment in all sectors was addressed by publication of the standard for environmental management systems, ISO 14001:2015.[2] Also, in 2015, the United Nations published their 17 Sustainable Development Goals and a *call for action by all countries – poor, rich and middle-income – to promote prosperity while protecting the planet.*[3] In 2016, the Building Research Establishment (BRE) published

---

[1] HM Government (2013) "Construction 2025". Crown Copyright.

[2] International Standards Organisation (2015) "ISO 14001:2015 Environmental management systems – Requirements with guidance for use".

[3] United Nations (2015) "Transforming our World: the 2030 Agenda for Sustainable Development". A/RES/70/1.

their technical manual to guide the adoption of sustainable materials and practices in International New Construction[4] and further advice for non-domestic buildings in the United Kingdom was issued in 2018.[5] References to BREEAM should also be applied to the establishment of strategic objectives. The UK Government produced the 10-point plan for a green industrial revolution in 2020 highlighting the need for greener and more energy efficient buildings. Sustainable construction is called for to support the environment by diminishing overall ecological footprints. Innovation is required in response to the ever-increasing demand for built space and the imperative to accelerate a circular economy that wastes fewer natural resources. Further information can be obtained from the following documents: The UK Climate Change Committee Carbon Budget[6] and the Dasgupta Report[7] (2020) and the IPCC Climate Change Report[8] (2021).

Turning to the need to improve performance, the Farmer Review (2016)[9] addressed the underlying causes of poorer than desired performance in the sector. The review highlighted the chronic problems associated with low margins, adversarial pricing models, financial fragility and overall poor image. Farmer described the imperatives to:

- address productivity and predictability

- deal with a lack of investment in innovation and training

- address the fragmentation of structures and associated leadership which results in poor collaboration between industry parties, and a culture, which is insufficiently committed to learning and continual improvement.

Dame Judith Hackitt led the inquiry into the fire at the Grenfell residential tower block in 2017[10] and a series of recommendations for building regulations and fire safety highlighting again the fragmentation of structures and leadership raised by Farmer (2016). Hackitt questioned the commitment to quality: addressing a 'culture of indifference' (2016:11) and a 'race to the bottom' (2016:5). Hackitt emphasised the need to address quality and the associated implications for building safety including improving a wide range of practices. These practices include but are not limited to:

- 'a greater appreciation of the voice of the end user' (2016:63)

- a focus on competence of parties engaged in projects (2016:73)

- creation and maintenance of a 'golden thread of building information' (2016:101)

- the accountability of duty holders for procuring quality goods and services representing good value, not necessarily lowest price (2016:108).

---

4   Building Research Establishment (2016) "International New Construction Technical Manual SD233". BRE Global.
5   Building Research Establishment (2018) "UK New Construction Non-Domestic Buildings Technical Manual SD5078". BRE Global.
6   Sixth Carbon Budget (2020) https://www.theccc.org.uk/publication/sixth-carbon-budget/.
7   The Economics of Biodiversity: The Dasgupta Review (2020) https://www.gov.uk/government/publications/final-report-the-economics-of-biodiversity-the-dasgupta-review.
8   Climate Change (2021) The Physical Science Basis. https://www.ipcc.ch/report/sixth-assessment-report-working-group-i/.
9   Farmer, M. (2016) "Modernise or Die, Farmer review of the UK Construction Labour Model". Construction Leadership Council.
10 Hackitt, J. (2018) "Building a safer future: independent review of building regulations and fire safety". Crown Copyright.

In 2009, the concept of the 'golden thread' was introduced to the industry in the Soft Landings Framework published by the Building Services Research and Information Association (BSRIA) and the Usable Buildings Trust (UBT). In 2012, the UK Government created Government Soft Landings (GSL) as a client requirement for project teams to deliver better projects utilising the Soft Landings concept. A 2018 update of the framework[11] publication reflects lessons learned by the industry. The aim of the Soft Landings approach is to influence the culture rather than deliver a different process model. When implemented on a project, the soft landings approach can deliver the changes to practices required by Hackitt including, but not limited to, the 'golden thread of building information'.

Creating the golden thread of building information is captured in the BS EN ISO 19650 standards[12,13,14,15] for managing information over the whole life of a built asset. This family of standards lays out the vision and requirements for information management to provide accessible, trustworthy, complete and accurate information about buildings and infrastructure.

Throughout this industry commentary, it is clear the notion of value is critical yet too often it is distorted and diminished by competing perceptions on where the value lies, particularly where collaborators are from different disciplines and in different tiers of the supply chain. In addition to defining value, it is important to have a clear understanding of the relative priorities of competing value-driven objectives. Value also needs to be procured and measured effectively. To address these matters, the Construction Innovation Hub have developed a Value Tool-kit that provides a way of defining, procuring and measuring value[16], emphasising the need for interaction, iteration and continuous review to ensure the desired social, environmental and economic outcomes from the project are delivered.

The UK Government has launched the Outsourcing Playbook (updated in 2020)[17], the Construction Playbook (2020)[18], and together with the UK Government Green Book (2020)[19]; they reinforce the need to consider a wide

---

[11] Agha-Hossein, M. (2018) "BG54 Soft Landings Framework: Six phases for better buildings". Building Services Research and Information Association.

[12] British Standards Institute (2018) "BS EN ISO 19650-1:2018: Organization and digitization of information about buildings and civil engineering works, including building information modelling (BIM) – Information management using building information modelling Part 1: Design phase of the assets". British Standards Institute.

[13] British Standards Institute (2018) "BS EN ISO 19650-2:2018: Organization and digitization of information about buildings and civil engineering works, including building information modelling (BIM) – Information management using building information modelling Part 2: Delivery phase of the assets". British Standards Institute.

[14] British Standards Institute (2020) "BS EN ISO 19650-3:2020: Organization and digitization of information about buildings and civil engineering works, including building information modelling (BIM) – Information management using building information modelling Part 3: Operational phase of the assets". British Standards Institute.

[15] British Standards Institute (2020) "BS EN ISO 19650-5:2020: Organization and digitization of information about buildings and civil engineering works, including building information modelling (BIM) – Information management using building information modelling Part 5: Security-minded approach to information management". British Standards Institute.

[16] Construction Innovation Hub (2020) "The Value Toolkit". Available at https://constructioninnovationhub.org.uk/value-toolkit (accessed May 2021).

[17] HM Government (2020) "The Outsourcing Playbook: Government Guidance on service delivery, including outsourcing, insourcing, mixed economy sourcing and contracting". Available at https://www.gov.uk/government/publications/the-sourcing-and-consultancy-playbooks (accessed 11 May 2021).

[18] HM Government (2020) "The Construction Playbook Government Guidance on sourcing and contracting public works projects and programmes". Available at https://www.gov.uk/government/publications/the-construction-playbook (accessed October 2021).

[19] HM Treasury (2020) "The Green Book: appraisal and evaluation in central government". Available at https://www.gov.uk/government/publications/the-green-book-appraisal-and-evaluation-in-central-governent (accessed 17 May 2021).

range of strategic and value-driven perspectives, alongside financial benefit to cost ratios, when preparing the business case for a project.

In this summary of strategic drivers, the content of the Construction Leadership Council's (CLC) 'Roadmap to Recovery' document (2020)[20], published to explicitly acknowledge the challenges posed by the Covid-19 pandemic (2020), is noted. The CLC make it clear recovery requires improved productivity (reduction of whole-life costs and increased delivery speeds), but these cannot be achieved without:

- competent individuals and firms, who

- design and build safe assets, to

- drastically reduce carbon emissions, and

- uphold fair practices.

To support the industry in addressing these strategic drivers in a practical way, the UK Government, as the largest client for construction in the United Kingdom, has produced the Construction Playbook 2020, with 14 mandated policies[21] to be used in the early life cycle to ensure the fundamental challenges inherent in sourcing and contracting public works are addressed. The Government are also actively endorsing sustainability through their support of the 2021 World Future Energy Summit and the United Nations Climate Change Conference of the Parties (CoP 26) in Glasgow with ambitions for a 'Green Industrial Revolution' through a 10-point plan of action.[22]

The case for change in how projects are managed in the built environment has multiple, complementary strategic drivers, and this edition of the *Code of Practice* is designed to highlight best practices and therefore address strategic drivers as far as practicable. This is done through adopting eight guiding principles to reinforce universal sector priorities, and eight themes applicable in specific ways across the life cycle.

Guiding principles

Guiding principles reflect the nature of contemporary project management in the built environment and the strategic drivers and priorities of the sector.

Eight principles serve to reinforce a holistic perspective when considering the management of all projects across the built environment, transcending silos and the specific priorities of individual stakeholders.

1. **Upholding the health, safety and well-being of people transcends all other objectives.**

   This principle applies to the people involved in creating the asset and those using the asset over time. The principle highlights the need for concern for the

---

[20] Construction Leadership Council (2020) "Roadmap to Recovery: an Industry Recovery Plan for the UK Construction Industry". Available at https://www.constructionleadershipcouncil.co.uk/news/construction-roadmap-to-recovery-plan-published (accessed 31 October 2020).

[21] HM Government (2020) "The Construction Playbook: Government Guidance on sourcing and contracting public works projects and programmes". Available at https://www.gov.uk/government/publications/the-construction-playbook (accessed December 2020).

[22] HM Government (2020) "The Ten Point Plan for a Green Industrial Revolution". Available at https://www.gov.uk/government/publications/the-ten-point-plan-for-a-green-industrial-revolution/title (accessed 17 May 2021).

health, safety and well-being of end users in addition to the duty of care the sector has to safeguard the project team members working in offices, on site, above and below ground, as well as members of the public affected by the works.

2. **Valuing diversity and promoting inclusion is a priority for all.**

Recent research, such as that published by the World Economic Forum,[23] demonstrates that diverse and inclusive teams bring greater creativity, innovation and profitability. Diverse teams contribute to higher performance through a broad range of skills, differing perspectives and better adaptability. People who value others for their skills and contributions rather than judging their characteristics are rewarded with higher performance and commitment.

In addition, international communities have initiated legislation, for example the UK Equality Act[24] ensuring discrimination against people on the basis of their personal characteristics, is unlawful. Protected personal characteristics include age, disability, gender reassignment, marriage and civil partnership, pregnancy and maternity/paternity, race, religion or belief, sex and sexual orientation.

Those who embrace the principles of diversity and inclusion are more likely to be successful and compliant with legal obligations.

3. **Ethical behaviours are non-negotiable, including and not limited to compliance with all relevant legislation.**

This principle emphasises the key requirement of any profession – for individual members to act ethically, beyond what is required by law, to do the right thing. Professionals must be trusted to work consistently in a moral, legal and socially responsible manner. CIOB reinforce this non-negotiable responsibility by requiring members of CIOB to[25]:

- act with integrity, honesty and trustworthiness,

- treat others with respect, fairness and equality,

- discharge their duties with complete fidelity and probity.

A summary of the core, specific, legislation relevant to projects in the built environment based upon English law is referenced in Appendix A of this *Code of Practice*. Every professional has a responsibility to be aware of the current version of legislation in the jurisdiction in which they are working. In addition, all professionals are expected to uphold the intent, as well as the detailed wording of all aspects of the law, for example upholding standards to enable data privacy and safeguarding of people at work or to prevent modern slavery and/or bribery and corruption.

4. **Projects in the built environment must contribute to achieving a net zero carbon and sustainable world.**

This principle addresses the strategic drivers associated with decarbonisation and sustainable construction from an environmental and socio-economic

---

[23] Esweran, V. (2019) "Business case for diversity in the workplace". Available at https://www.weforum.org/agenda/2019/04/business-case-for-diversity-in-the-workplace (accessed 20 November 2020).

[24] The Equality Act, 2010

[25] Chartered Institute of Building (2018) "Rules and Regulations of Professional Competence and Conduct". Available at https://uat-olaersy-obn5vwqw2ewzg.uk-1.platformsh.site/sites/default/files/2021-05/Rules%20and%20Regulation%20of%20Professional%20Competence%20and%20Conduct.pdf (accessed 25 May 2021).

perspective. It addresses both the what and the how of sustainability in the built environment. From the perspective of environmental impact, sustainable construction is concerned with the design and management of built structures, the performance of the selected materials, and the adoption/ use of renewable energy resources and aligned technologies to reduce global greenhouse gas emissions. From the perspective of economic impact, sustainable construction is concerned with the transition to a circular economy including renewable energy generation, material and waste recycling, and water harvesting and preservation.

5. **Successful projects have a whole-life perspective and are developed with the needs of end users at the forefront.**

This principle supports the approach taken in the *Code of Practice* to focus on the asset life cycle and the impact of projects across all stages of the life cycle. 'Beginning with the end in mind', i.e. considering how an asset will be used, maintained and eventually retired, is a vital part of project management owned and led by the client.

6. **Project management is a strategic enabler addressing business opportunities and focused on creating benefit.**

This principle highlights the role of project management in creating assets to provide quality solutions to a current challenge at an appropriate pace and for a whole-life cost affordable to the client. Alongside these competing priorities of quality, time and cost, project management should endeavour to ensure the impact on people and the planet is positive and likely to remain positive over the life of the asset. The strategic nature of projects requires long-term and systemic thinking, and a focus on balancing needs with affordability, achievability and adaptability to respond to uncertainty and change.

7. **Focus is required on both physical and information assets across the life cycle.**

This principle accentuates the requirements to deliver quality information to support commissioning, handover, use, maintenance and disposal of the asset alongside the planned physical deliverables. Information may be structured (e.g. tables of equipment/components and their priorities) or unstructured (e.g. PDF's of performance certificates). It will include digital assets including, but not limited to, building information modelling (BIM)[26], and other assets required for value to be created for clients such as software and digital monitoring and performance management tools. These create the Golden Thread of knowledge for the asset.

8. **Projects are complex and leadership is needed to deal with issues relating to people, politics, power, conflict, uncertainty and disruption.**

This principle highlights that complexities associated with projects in the built environment are not limited to spatial, technical and/or physical matters. The context for projects is often complex. Multiple stakeholders hold competing objectives and there can be significant uncertainty associated with emergent risks and disruptive trends. Further, complex supply chains

---

[26] At the time of publication, it is correct to refer to building information modelling (BIM) as part of overall information management, although increasingly emphasis is on using the latter term.

Introduction

can cause competition and the erosion of value rather than collaboration. Leadership of projects across these domains is vital for success.

Themes

The CIOB acknowledge that for projects and programmes of work, it is imperative the project manager meets the standards and expectations informed by the clients' requirements. The project manager must adopt best practice to manage the planning of all processes and procedures related to the project. This will include, but not be limited to, the following:

- a fully coordinated design

- fully validated assurance, including HSW and quality, and quality control procedures

- a fully coordinated master plan which is regularly updated and issued to all relevant stakeholders

- awareness of, and implementation of, all innovations and client/Government objectives, for example environmental improvement

To uphold the guiding principles and to reflect the strategic drivers, eight **themes** are applied across the life cycle stages.

All are linked but the eight themes chosen highlight specific, contemporary, priorities for project management in the built environment as follows:

A. **Quality**: addressing the need to clearly define standards and the associated requirements of clients and end users including, but not limited to, quality and building safety and to ensure and satisfy specifications correctly, the first time.

B. **Sustainability**: addressing the need to be stewards of the natural world, addressing air quality, climate change, water usage, land usage, resource usage and biodiversity and creating an effective circular economy, to step away from a paradigm of 'make, use, dispose'.

C. **Value**: addressing the socio-economic value of the asset in terms of whole-life costs and benefits and ensuring there is a competent process to decide on the relative priorities of drivers of value including economic and social objectives, to procure and measure performance to validate priorities.

D. **Productivity**: addressing the need to innovate and use appropriate modern methods of construction to optimise life cycle cost and profitability, managing the pace of build and the achievement of quality and sustainability targets.

E. **Leadership**: addressing the need to focus on people, their skills, well-being and career opportunities, and also the need to provide competent governance and decision-making to lead the team through uncertainty and complexity.

F. **Collaboration**: addressing the need to build strong relationships across diverse networks of people and organisations, upholding inclusivity and equality of opportunities for work across the asset life cycle.

G. **Knowledge**: addressing the need to create and control the use of knowledge, including intellectual property, to produce complete and accurate information to support the asset in use, over the whole life cycle, and to ensure a culture of sharing, learning and continuous improvement.

H. **Risk**: addressing the need to explicitly deal with uncertainty and ambiguity, to improve predictability and the ability to meet objectives. To ensure value is protected by an appropriate mix of identification, elimination, reduction of risks and preventative controls and/or contingency, business continuity and resilience plans.

Themes are woven through the structure of the remainder of the *Code of Practice*, highlighting how these themes apply to each life cycle stage.

Structure of the *Code of Practice*

The remaining eight chapters of the *Code of Practice* are organised around the stages of the asset life cycle as follows (and as shown in Figure 0.1):

1. **Identify**: needs and benefits

2. **Assess**: options and feasibility

3. **Define**: approach and procurement strategy

4. **Design**: specifications and functionality

5. **Implement**: manufacture and construction

6. **Validate**: integrate and handover

7. **Operate**: use and maintain

8. **Retire**: re-purpose or demolish

The successful completion of any project or programmes of work requires the project manager to have a solid awareness of the contents of each chapter to ensure the project plan can be understood, updated, implemented and monitored.

Each chapter addresses the:

- **Purpose** of the life cycle stage

- Applicability of each **Theme**

- **Activities** required to meet the purpose and apply the Themes

- **Key roles and responsibilities** for each activity

- **Decisions** required to complete the stage

Roles and responsibilities are focused on six generic roles as shown in Table 0.3 to reflect the client perspective across the whole life of the asset. Professionals who use this *Code of Practice* can also include representatives of the supply chain, including consultants or sub-contractors who are likely to define additional roles and responsibilities for specific sub-sets of the project; however, these should always reflect the overall client context.

References to directly relevant external publications, other CIOB publications and Guidance Notes (see Appendix) are made in the form of footnotes.

Introduction

Table 0.3 Roles and responsibilities for six generic roles

Role	Description
End users	Occupants or users of the built environment.
Operator	Responsible for operation and maintenance of the asset as designed and built on behalf of the client and in compliance with all relevant legislation. *In some situations, the operator/ maintainer may be the same entity as the end user/occupier.*
Client Sponsor	Accountable, on behalf of the wider client organisation, for achieving beneficial outcomes from the project including representing the needs of end users and funding bodies.
Client Project Manager	Responsible to the client sponsor for achieving project objectives. The project manager may be a client employee or consultant. In either case the project manager ensures the administration of any contract(s) on behalf of the client. The Client Project Manager should assist with support where necessary in developing a client's strategic brief, which should include CDM, establishing the project and design teams. Accountabilities should be validated and informed by the use of RACI principles through all project stages.
Consultant	Specialist advisors to the client team, for example architects, engineers, technology and process/method experts. Consultants may appoint their own responsible person who reports to the client project manager for the contracted scope of work. This person may be called the consultant's project manager. It is imperative all scopes of work for consultants are defined with specific roles and responsibilities prior to commencement of any design.
Contractor	Responsible for delivering the design, build or maintenance of the physical asset, in whole, or in part, in line with the contract(s) administered by the client project manager. Contractors may appoint their own responsible person who reports to the client project manager for the contracted scope of work. This person may be called the contractor's project manager. It is imperative all contracts are defined with specific roles and responsibilities prior to commencement of any outputs.

Note: Roles and responsibilities from the client perspective do not in any way infer nor supersede legal liabilities, for example the Construction (Design and Management) Regulations, 2015 (CDM, 2015), clearly define the responsibilities of the Principal Designer who could be a client employee, or ideally the lead designer working for a design practice or consultant and/or passed onto the principal contractor under a design and build contract, and where details will be defined in the project execution plan.

References are limited to authoritative sources such as ISO standards and widely applicable sector reports. Books, academic publications and other sources are excluded.

Application to practice

Projects in the built environment span a wide range of scenarios, different assets, different funders, different scales and complexities.

Examples are used throughout the text to illustrate the breadth of application and to highlight particular common situations and areas to consider when applying the advice in this *Code of Practice* to projects.

Application to practice clearly requires a working knowledge of relevant legislation. It is impossible to cover all core, specific legislation in the built environment given the diverse nature of assets and infrastructure settings. Readers are directed to the legislation advised in the British Standards Institute work on the overarching framework for competence of individuals in the built environment (see Table 0.4). This framework outlines the requirement for *awareness of and contribution to compliance with all relevant requirements of building regulations for public health and public safety* (2020:12) and specifically notes the legislation relevant to the roles of principal designer, principal contractor and

Table 0.4 Core and specific legislation relevant to projects in the built environment

Health and Safety at Work etc. Act, 1974
Construction Design and Management (CDM) Regulations 2015
The Housing Act 1985, 1988, 1996, 2004
Town and Country Planning Act 1990
Housing, Construction and Regeneration Act 1996, as amended 2011
Dangerous Substances and Explosive Atmospheres Regulations 2002
Building Safety Bill 2021 as appropriate at time of publication
Fire Safety Act 2021

building safety manager in higher-risk buildings.[27] Guidance Notes providing an overview of each area of legislation in the table are included in Appendix A.

Other references that can be considered, include*:

- (Draft) Building and Buildings, England: The Higher-Risk Buildings (Descriptions and Supplementary Provisions) Regulations [2021][28]

- (Draft) Building and Buildings, England: The Building (Appointment of Persons, Industry Competence and Dutyholders) (England) Regulations [2021][29]

- (Draft) PAS 9980: Fire risk appraisal and assessment of external wall construction and cladding of existing blocks of flats – Code of practice[30]

*Note, the above are, at the time of writing, drafts and not yet incorporated into legislation or Standards.

Readers are also guided to the CIC 2020 publication 'Setting the Bar' (2020) focused on a competence regime for building a safer future.[31]

It is the responsibility of professionals in all jurisdictions to be aware of the detail of the current version of applicable legislation where they practice.

GUIDANCE NOTES in the Appendix to the *Code of Practice*

The Appendix contains a number of Guidance Notes providing a more detailed description of a particular activity, relevant legislation or practice. Where the chapters focus on the 'why', 'what', and 'who' for each stage, guidance notes provide more detail on the 'how'.

Guidance notes may not be universally applicable to all projects in the built environment but are included where there is wide usage of a particular concept, technique, practice or piece of core legislation.

---

[27] British Standards Institute (2020). "Built environment – Overarching framework for competence of individuals – Specification". BSI Flex 8670: v1.0 2020-09.

[28] https://www.gov.uk/government/publications/building-safety-bill-draft-regulations

[29] https://www.gov.uk/government/publications/building-safety-bill-draft-regulations

[30] https://standardsdevelopment.bsigroup.com/projects/2020-01838#/section

[31] Construction Innovation Council (2020) "Setting the Bar". Available at https://cic.org.uk/admin/resources/setting-the-bar-9-final-1.pdf (accessed 20 May 2021).

In summary

Effective project management is central to the creation and use of the built environment in a way that delivers value to society and meaningful work to millions of people around the world.

Construction is transient; hence, the project manager is of critical importance related to the robust planning and management of the project or programmes of work.

Since the fifth edition of the *Code of Practice,* there have been significant advances in the agreement of strategic drivers and priorities for the sector. Historic practices are challenged as thinking has advanced as a result of experience. There is clearly much about project management, and much about the context of the built environment, which has not changed; however, the sixth edition incorporates significant changes to embed strategic drivers and priorities for the sector as core guidance.

The remainder of Chapters 1–8 focus on each stage of the asset life cycle in turn, providing clear advice for practitioners involved in the management of projects across the built environment. The Appendix provides guidance on specific activities, practices and legislation relevant to planning and managing projects across the asset life cycle.

1 Identify: needs and benefits

Purpose

> The purpose of the **Identify** stage of the life cycle is to ensure client needs are clearly understood and the high-level benefits for end users, society and funders are defined in measurable terms, documented and agreed by the client.

The eight Themes are applied in the Identify stage as listed below. It is recommended that information about their application is documented and shared with known project stakeholders, in an easily accessible repository.

- Agree principles and strategies for **quality**, including building safety. These should be based on client expectations and made available to known project stakeholders.

- Agree aims for **sustainability** based on legislation and client expectations.

- Agree desired socio-economic **values** ensuring the process of agreeing the value drivers with the client is documented.

- Agree strategy for innovation and **productivity** taking account of the market and available technologies, and ensure decisions are documented.

- Establish governance to create the environment for effective **leadership**, agree terms of reference and roles and responsibilities.

- Agree principles for **collaboration** and risk-sharing between the client and all contractors and consultants in the supply chain.

- Agree the purpose and scope of information as part of wider **knowledge** management.

- Express appetite and capacity for **risk** in measurable terms and identify risks to strategic, regulatory, commercial and reputational objectives. Establish a risk register.

Activities

The purpose and application of the themes is achieved through implementation of the following activities and creation of documented deliverables.

The list of activities does not represent a strict sequence of work but is indicative of a likely order. In practice, many activities will happen in parallel and with

Code of Practice for Project Management for the Built Environment, Sixth Edition. Chartered Institute of Building.
© 2022 John Wiley & Sons Ltd. Published 2022 by John Wiley & Sons Ltd.

iteration in order to create the deliverables needed to approve progression to the next stage of the life cycle.

A	• Agree business opportunity and funding routes
B	• Confirm sponsor and governance arrangements
C	• Confirm client project manager
D	• Confirm key stakeholders and needs/high-level requirements
E	• Confirm any existing assets in scope
F	• Define measurable benefits
G	• Establish the information management approach
H	• Review past experiences and visit relevant benchmarks
I	• Document and approve project mandate
J	• Document and approve high-level business case
K	• Document and approve initial stakeholder analysis

A. **Agree business opportunity and funding routes**. The client is accountable and responsible for agreeing the business opportunities and identifying the available funding. The link between the organisational strategy or policy commitment and the statement of the business opportunities should be clear and reflective of the organisational context including any emergent risks or disruptive trends, for example to reflect organisational environmental policies including decarbonisation or recycling targets. In some contexts, the business opportunity and funding routes will be managed through a portfolio of projects, or a master programme. For example, a collection of projects associated with refurbishing assets within a water utility may be managed as a portfolio to ensure efficient use of resources, or a collection of projects to build new hospitals may be managed as a master programme to enable common design and build standards to be maintained and to ensure funding is targeted as a priority. Depending on the context, funding may be solely by the client, from reserves or borrowing, or may involve a consortium of funders and funding instruments.[1] An in-principle agreement on the opportunities and funding strategies, typically made by an investment or executive committee in the 'business as usual' organisation, triggers the project and the remaining activities in this stage.

B. **Confirm sponsor and governance arrangements**. The ultimate client stakeholder is accountable and responsible for appointing a person to the role of **client sponsor** and for agreeing governance arrangements, including (1) the organisation's appetite for risk[2] in relation to the project; (2) translation of the appetite for risk into delegated limits of authority for the client sponsor; (3) escalation routes for matters breaching delegated limits; and (4) assurance mechanisms to provide confidence to the client that the project is in control.[3] The client sponsor represents the client and has the delegated authority to make decisions on behalf of the organisation. Where the project

---

[1] Guidance Note 1: Funding mechanisms
[2] Guidance Note 2: Risk appetite and delegated limits of authority
[3] Guidance Note 3: Assurance and the three lines model

will be funded by a single privately owned organisation, confirming governance arrangements may be relatively simple. Where the project has multiple funders or has public-sector investment, establishing sponsorship and governance will be more complex and take time to establish. Governance with associated terms of reference needs to be specified for the Identify and Assess stages of the life cycle. Governance arrangements for the stages of the life cycle where there will be significant capital investment are defined in the Define stage. These may be adjusted as the asset enters the Operate stage of the life cycle.

C. **Confirm client project manager**. The **client project manager** is the person who will take day-to-day responsibility for ensuring the project is planned and controlled in line with the ultimate client stakeholder requirements and governance arrangements. The client may appoint a member of staff or appoint a **consultant** or **contractor** to undertake this role. Where a consultant or contractor is chosen, the client project manager role must not be confused with other project management roles established to take responsibility for any part of the scope of work in line with a contract or contracts. It can be common for multiple people to have the title of project manager as part of a large project, for example the main contractor's project manager, or a consultant's project manager; however, it is the **client project manager** who has responsibility throughout the project life cycle, administering any contracts associated with the project through to the cessation of commissioning and contract end, including any claims or dispute resolution. This will extend beyond the physical handover of the asset to the **operator**. Depending on the complexity of the project or framework of projects, there are multiple options for the contractual procurement route and design of the project organisation and the use of internal and external expertise across the life cycle.[4]

D. **Confirm key stakeholders and needs/high-level requirements**. The **client sponsor** is responsible for ensuring key stakeholders are identified, including representative **end users** and **operators** who must provide input to the definition of needs and how these are expressed as high-level requirements for the project. The client's project manager appointment should include responsibilities to assist with the identification and management of stakeholder engagement. Where there is no opportunity to engage representative end users or the operator, the client sponsor should seek advice from third parties with relevant expertise, for example a champion with specialist facilities management expertise. It is vital at this stage representative end users and other influential external stakeholders have the opportunity to influence thinking. Depending on the asset to be built, refurbished or improved, this will include planning authorities at local and potentially national levels and regulators and related stakeholders relevant to consenting beyond planning approvals.[5] Stakeholder analysis and mapping[6] needs to ensure all individuals or groups with the potential to help or challenge the project are considered. These may include local communities, heritage and

---

[4] Guidance Note 4: Design of the project organisation
[5] Guidance Note 5: Consenting considerations
[6] Guidance Note 6: Stakeholder analysis and mapping

1 Identify

conservation bodies, tenant associations, elected politicians and a range of other interested parties. Understanding the needs/high-level requirements of key stakeholders must include a consideration of each Theme to ensure the project accounts for all relevant drivers of value early: the quality principles to be applied, aims for sustainability, the socio-economic value required, the drivers to innovate, the need for productivity, lessons to be learned from previous projects and experiences, and how leadership will ensure appropriate collaboration and risk-taking across the supply chain.

E. **Confirm any existing assets in scope.** The **client project manager** is responsible for confirming assumptions about the high-level scope of the project. This will include consideration of which existing assets are in scope for this project. Some projects may be associated with 'green-field' sites. Projects may involve a mix of works to demolish and/or dispose, re-cycle and/or re-purpose, as well as creating new assets, for example a project to create a new school may need to demolish buildings on an existing site before or after new facilities can be built, as well as re-purpose other existing assets. All existing assets in scope need to be identified, with the **operator** of existing assets providing essential input. Engagement with external stakeholders with authority with respect to listed buildings, heritage sites and associated planning conditions is important at this stage to validate assumptions to be included in the high-level business case.

F. **Define measurable benefits.** Investments in projects in the built environment are justified by the benefits they create for the client, their funders, the end users of the asset and the wider society in which the built asset or infrastructure will exist. Benefits are the measurable improvement resulting from achievement of an outcome – these are the quantified expression of value created by the project, for example financial return, improved satisfaction with living conditions, reduced carbon footprint or contribution to a government policy to provide jobs in a certain region. Robust quantification of benefits is required to ensure the business case for the project does not mislead decision-makers and key to this activity is mapping the route to benefits[7], defining the measures to be used to establish the baseline and tracking the realisation of benefits over time.[8] This is critical at outset as it would be wasteful to continue to work on the project if there was no business case, and because benefits data should be available from the operator of any existing assets in scope. The accountability for this activity is with the **client sponsor**; however, they may delegate responsibility to the **client project manager** to work closely with the **operator** and **end users** where possible, and to provide the information to the client sponsor for approval. The **client project manager** may appoint specialist consultants to assist with this work.

G. **Establish the information management approach.** The **client sponsor**, along with the need to confirm high-level requirements, is responsible for defining the purpose and scope of information management on behalf of all project stakeholders. The **end user and/or operator** is a key stakeholder associated with this activity as they will be accountable for the realisation of project benefits in the Operate stage. Considering Soft Landings[9], and

---

[7] Guidance Note 7: Benefit mapping
[8] Guidance Note 8: Benefit measurement and realisation
[9] Guidance Note 9: Soft landings framework

guided by the BS EN ISO 19650 series of standards[10], the **client sponsor** must make the purpose and scope of information management clear at this early stage of the project to ensure adequate provision is made for resourcing this need, as well as for the physical creation of the asset. This will include ensuring all appointed consultants are aware of all decisions relevant to their scope of work. BS EN ISO 19650-2 includes a series of activities to be completed by the client before any project-specific appointments are made to ensure the client is ready to oversee the information management process and has the appropriate documentation, tools and processes in place. Establishing the information management approach is wide ranging and includes establishing or reviewing all corporate or project-wide information management resources. These include organisational, asset and project information requirements (the OIR, AIR and PIR), as well as the project information standards, existing reference information and the project technology solutions for managing and retaining information. For example, existing information for a new building on a site already occupied by the client could include the site boundary plan and the ground investigation reports from previous projects.

H. **Review past experiences and visit relevant benchmarks.** In the Built Environment, few projects are entirely unique and learning from others is critical. The client may be a mature project organisation, such as one funded by a government to manage a strategic road network, or a private limited company involved in house building. Here, the organisation should have a wealth of knowledge to draw upon when shaping new projects. Less mature clients, for example a SME (Small-Medium enterprise) embarking on creating a manufacturing facility, for the first time may not have this knowledge internally but will benefit from engaging with other bodies who can help them to learn and avoid common pitfalls. This applies to the design and construction, and also the operation, maintenance and eventual retirement of assets.

I. **Document and approve project mandate.** The **client project manager** is responsible for documenting the project mandate[11], bringing together information on the business opportunity, funding, governance arrangements, the expression of needs and high-level requirements, the information purpose and scope, information about existing assets in scope and the high-level benefits of the project. The project mandate is a key input to the high-level business case and provides the single description of the project. At this stage of the project, the project mandate should outline any assumptions, planning approvals forthcoming or timeframes, for example the need to meet a particular opening date for a school, shopping mall or sporting venue. The project mandate is one of the critical deliverables to be approved by the **client sponsor** and wider governance at the stage-gate review, which triggers approval to move to the Assess stage. It forms the basis for the *project brief* in the next stage, and in time the *project execution plan*.

J. **Document and approve high-level business case.** The high-level business case should include a full appraisal and analysis of the various potential options with detailed recommendations. This document should be distributed and approved by all relevant project stakeholders. The approved high-level

---

[10] Guidance Note 10: Information management based on the BS EN ISO 19650 series of standards
[11] Guidance Note 11: Project mandate indicative content

business case is the formal document used to initiate the project delivery and should provide recommendations for procurement and contractual arrangements. The responsibility for this activity is with the **client sponsor**.

The term 'high-level' business case is used here; however, there may be alternative descriptors depending on the client organisation, for example strategic outline business case (SOBC). A business case needs to bring together the rationale and justification for the project. The financial case (investment appraisal[12]) determines affordability and the return on investment by analysing benefits, whole-life costs and associated risks. Non-financial aspects of the business case also need to be included, for example benefits such as reduced carbon footprint and/or social impact will be measured in non-financial terms but are nevertheless part of the internal and external justification for investment.[13]

Typical risks relevant at this stage of the project may include land ownership or feasibility of envisaged construction methods. At this stage of the project, it would be typical for estimates of benefits and costs to be 'order of magnitude' (class 5) estimates.[14] The level of definition would be approximately 10% and with a typical accuracy of −50/+100%. The client project manager may engage specialist consultants, for example architects or quantity surveyors to support the estimation process. Client organisations and funders will have different requirements for preparation of the business case and practitioners must align their work with the client requirements.

K. **Document and approve initial stakeholder analysis.** The **client project manager** is responsible for capturing known information regarding stakeholders, their opinions and anticipated influence and involvement in the project. It is imperative clear, concise and manageable terms of reference are established and approved to clarify the roles and responsibilities of all project stakeholders. Many techniques may be used to capture and represent such information, for example using a matrix or flow or bubble charts to illustrate likely stakeholder influence relative to their interest in how/when the asset is built, or a graphic to represent the network/connections between each identified stakeholder (see Guidance Note 6). The purpose is to be clear, before approval to move to the Assess stage, as to the individuals and groups who may have a significant impact on the project because (1) they are influential and may impact decisions and progress and (2) their views and behaviours may impact the realisation of benefits. This is a collaborative activity and the **client project manager** is responsible for capturing the views of representative **end users**, **operator(s)**, funders, regulators and influential internal and external groups. Specific stakeholders will clearly vary depending on the project and client in question. For example, in a complex, multi-investor infrastructure project such as a major rail or road scheme, stakeholder analysis will consider parties impacted by compulsory purchase orders and the political agenda of local councillors or members of parliament. Whereas a project concerned with refurbishing a retail outlet housed within a listed building will consider parties with regulatory influence, as well as brand and service-related requirements of the client.

---

[12] Guidance Note 12: Investment appraisal
[13] Guidance Note 13: Business case
[14] Guidance Note 14: Estimating

Decisions

Before approving the project to move on from this stage, the defined decision-makers and the governance board, are required to confirm:

1. Needs of key stakeholders are clearly defined in writing to enable clarity of understanding between the participants of the vision and high-level strategy for the development and use of the asset. For some organisations this information may be informed by definition of a wider portfolio or programme.

2. Any existing assets to be modified or retired by this new project – whether in whole or in part, with a full appraisal or detailed recommendations prepared by the project manager.

3. The purpose and scope of information required to support the development and use of the physical asset.

4. Client sponsorship and governance are in place. This should include a clear organogram of governance levels and hierarchy, combined with a responsibility matrix.

5. Funding is in place for the next stage with a clear line of sight of how the whole project will be funded. The funding documentation, including heads of terms between contracted entities, should be clearly defined and agreed.

6. Known external dependencies, constraints and risks in the wider context, should be listed, defined, validated and represented in the high-level business case, including matters related to planning consents.

7. Deliverables adequately address each of the defined Themes.

   (a) The principles and strategy for **quality**, including building safety, are clear and agreed and incorporated into the Project Execution Plan.

   (b) The high-level impact of the project on the environment is understood and **sustainability** aims and goals are clear, agreed, recorded and meet with national and local governmental and corporate objectives.

   (c) The desired socio-economic **value** is agreed and recorded.

   (d) Innovation strategy and **productivity** requirements are agreed and recorded, taking account of the market and available technologies, the project manager should be aware of future potential innovations and their impact on the programme of work.

   (e) Structure and approved governance will enable the context and culture for effective **leadership.**

   (f) The principles for **collaboration** and risk-sharing between the client and all consultants and contractors and the wider supply chain are clear, agreed and documented.

   (g) The purpose and scope of information management as part of wider **knowledge** management are clear, agreed and documented.

   (h) The appetite and capacity for risk is expressed in measurable terms. Risks to strategic, regulatory, commercial and reputational objectives are identified and documented.

1 Identify

The scope of work for the Assess stage is understood and a suitably competent and experienced team is in place to do this work (see Guidance Note 4), for example consultants who can support environmental impact assessments or representatives of the operator to ensure the chosen solution sufficiently consider operability, maintenance and whole-life costings.

2

Assess: options and feasibility

Purpose

The purpose of the **Assess** stage of the life cycle is to ensure the feasibility of a range of options are assessed and evaluated to meet the need and benefits and are within the appetite for risk. The chosen option for the asset is approved by the client as being the most suitable in the circumstances. It is essential the project manager applies a robust and logic-driven process to manage all stages of the life cycle. Each stage of the life cycle will have interdependencies and any failure to fully understand and regularly update this process may lead to delays, some of which may not be recoverable.

The eight Themes are applied in the Assess stage as follows:

- Define **quality** criteria and priorities based on client expectations.

- Define **sustainability** design principles based on legislative and client requirements.

- Define **value** drivers and priorities based on client expectations.

- Define all assumptions regarding **productivity**, for example the mix between on-site and off-site construction. These choices will need to be analysed and reviewed as several options may be feasible; however, time, quality and value requirements will influence the chosen method.

- Maintain a **leadership** focus on long-term principles and objectives rather than short-term gain, ensuring a focus on team structure and terms of reference.

- Consider **collaboration** of stakeholders and the supply chain for each option by promoting early involvement.

- Share **knowledge** and capture information to support the selection of the chosen option and ensure the data is stored on an accessible repository.

- Ensure **risks**, threats and opportunities and associated contingencies are assessed for each option and recorded and updated at regular intervals within documentation accessible to stakeholders.

Activities

The purpose and application of the themes is achieved through implementation of the following activities and creation of documented deliverables.

Code of Practice for Project Management for the Built Environment, Sixth Edition. Chartered Institute of Building.
© 2022 John Wiley & Sons Ltd. Published 2022 by John Wiley & Sons Ltd.

2 Assess

The list of activities does not represent a strict sequence of work but is indicative of a likely order. In practice, many activities will happen in parallel and with iteration in order to create the deliverables needed to approve progression to the next stage of the life cycle.

A	• Establish the team and governance for the Assess stage
B	• Define decision criteria for option selection
C	• Identify and analyse options
D	• Select and acquire site
E	• Engage key stakeholders and preferred option(s)
F	• Choose best fit option and document the decision
G	• Document and approve project brief
H	• Document and approve intermediate business case
I	• Document and approve stakeholder map

A. **Establish the team and governance for the Assess stage**. The **client sponsor** is responsible for commissioning a team of people, which is suitably competent and experienced, to identify options to meet the needs and benefits and to analyse the feasibility of each option using a rational approach. The team must include people and organisations with the appropriate information management capabilities and capacity, who are provided with a specific set of information requirements by the client (see Guidance Note 10). Specialists from consultancies or contracting organisations may be commissioned to support this work. In addition to core skills in designing, surveying, procuring, construction or project planning, experts with specialities in, for example, consenting, sustainability, information management or a particular regulatory requirement may be required. The **client project manager** is responsible for bringing the team together, with responsibility for confirming terms of reference, establishing effective working practices and leading the team to complete the activities in this stage to the satisfaction of the client sponsor and wider governance.

B. **Define decision criteria for option selection**. The **client project manager** is responsible for establishing and documenting a clear view of the decision criteria from the client perspective. In all cases requirements associated with quality, sustainability[1], value and productivity will need to be translated into clear decision criteria to fit the tolerances of risk appetite and capacity for risk (Guidance Note 2), and the desired benefits (Guidance Note 7) developed in the Identify stage.

Dependent upon the asset to be developed, there will be different considerations to address as part of the decision criteria and decision-making process. When considering future maintenance, in terms of access, space and product choice, there will be different drivers; a retail outlet may consider disruptions to footfall, a housing development, the need to access private spaces, a road or rail scheme, disruption to the network. There may be requirements or targets for how much disruption to natural habitat may be tolerated, requirements to gain permits for the disposal of site waste materials or the desire to establish a 'green lease' arrangement with the

---

[1] Guidance Note 15: Materials selection

future operator to ensure a focus on sustainability across the life of the asset. Use of one of the sustainability assessment schemes, BREEAM[2], LEED[3], CEEQUAL[4], may be required. Considerations of social impact may be the proximity of a school or hospital to housing, the air quality a local community would tolerate to introduce industry and employment or trade-offs regarding brand signage if the chosen site has heritage limitations. These examples demonstrate the need for decision criteria to address the priorities, needs and benefits brought forward from the Identify stage.

Decision criteria must also be reflected in the information requirements and the client project manager must document the detailed information to be delivered by each internal and external member of the project team appointed to identify options – the 'exchange information requirements' in ISO 19650-2 (see Guidance Note 10). The **client project manager** is responsible for documenting each decision criterion and gaining agreement of the decision-making process[5] from the client sponsor and wider governance.

C. **Identify and analyse options**. The team, led by the **client project manager**, identify options for consideration, addressing the decision criteria and meeting the overall needs and benefits defined in the Identify stage. It is important the process of identifying options is creative and challenges thinking on different ways to achieve the needs and benefits expressed by the decision criteria. The generation of information by each party involved must be consistent with the client's exchange information request, for example a water treatment project may need to address population change, types of industry and employment and opportunities for reducing carbon emissions from treatment processes. Each project is unique, and the degree to which tried and tested options are valuable, in comparison with innovative or unfamiliar options, will vary in practice. This is most relevant if the client has committed to quality, sustainability, value and/or productivity priorities that are more demanding than on previous projects. When the team has identified distinct options, these can be analysed using decision criteria and the approved decision-making process. Good practice in design management is critical to success.[6] At this stage, high-level options including, for example, renovation, enlargement, demolition or new build could be analysed via cost estimation to assess initial viability.

D. **Select and acquire site.** Site selection and acquisition is an important stage in the project cycle where the client does not own the site to be developed. This should be addressed as early as possible and, ideally, in parallel with the feasibility study. (It is to be noted that the credibility of the feasibility study will depend on the major site constraints.) A specialist consultant and legal advice may be required and may involve substantive due diligence, and this must be managed by the client project manager. The site requirements must be defined in terms of the asset to be constructed and must be acquired within the constraints of the outline development schedule, and with minimal risk to the client. In some circumstances, the client may already own the site; however, if this is not the case, site selection and acquisition

---

2 Assess

is critical.[7] The consideration of potential sites should be carried out in parallel with the assessment and selection of the concept for the project. The **client project manager** is responsible for managing this work and making clear to decision makers the status of land/site acquisition and any associated consents required, for example, the need for an environmental impact assessment (EIA). Critical information is provided by site/ground investigations[8]; boundary issues and any third-party connected issues must also be reviewed and managed by the client project manager to avoid land-related problems during the construction phase. A boundary review may be particularly relevant to verify and validate actual and legally approved site boundaries for older properties where historical documentation may be hand-written.

E. **Engage key stakeholders on preferred option(s)**. The preferred option will not only satisfy specific decision criteria relating to quality, sustainability, value and productivity, but should ideally be supported by the key stakeholders determined in the Identify stage, for example, local community representatives, regulatory bodies and/or elected politicians (see Guidance Note 6). The **client project manager** is responsible for acquiring inputs from key stakeholders on the preferred option. Key stakeholders should include representative end users, the client, operator, key regulatory bodies (for example planning and local authorities), consultants (for example lead designer) and supply chain representatives, contractors and suppliers. This does not infer contractors have been selected; however, the market confirms the chosen option is achievable. The process for gaining input from key stakeholders will vary depending on the project, for example focus groups could be used with representative end users, and/or a series of individual consultations could take place with different players from planning authorities, heritage organisations or community groups. The objective is to ensure stakeholder voices are accounted for and to relay the commitment to action the chosen option. Stakeholders, whose governance levels are defined on the requisite authority matrices, may have the power to veto any decision, but understanding of perspectives and commitment to action is nevertheless critical to the future successful execution of the project.

F. **Choose best-fit option and document the decision**. The **client sponsor** is responsible for the decision of the chosen option, as specified by project governance. The **client project manager** validates the rationale for the decisions and is responsible for documenting both the process and decision, providing a comprehensive and transparent audit trail and a clear description to inform the project Definition and Design stages. The identification of milestones, gates or stages, within the project master plan and highlighting the need for sign-off by the client sponsor is fundamental.

G. **Document and approve project brief.** The **client project manager** is responsible for documenting the project brief, taking information from the project mandate and expanding this to include information about the chosen option. The project brief is a key input to the intermediate business case providing a clear description of the asset(s) to be delivered[9] and any associated

---

[7] Guidance Note 18: Site selection and acquisition

[8] Guidance Note 19: Site investigations

[9] Guidance Note 20: Project Brief indicative content

requirements, for example, safety and environmental targets and minimum standards. The brief provides clear expectations for the early appointment of consultants, designers and/or contractors. The project brief will update any assumptions or known information about timeframes and external dependencies and constraints, for example, operational parameters such as weather-proofing or functional constraints such as height limitations. The project brief is a critical deliverable and is approved by the **client sponsor** and wider governance at the stage-gate review. This triggers approval to move to the Develop stage. After the Assess stage, the project brief becomes incorporated into the Project Execution Plan (PEP). For some clients, a Project Initiation Document (PID) is used as an alternative to the project brief, and sometimes as an alternative to the project execution plan (see Guidance Note 30).

H. **Document and approve intermediate business case.** The responsibility for this activity remains with the **client sponsor**, however they may delegate the work to the **client project manager** to create the second iteration of the business case (Guidance Note 13) and agree this with the **client sponsor** and wider governance. The term 'intermediate' business case is used here, but this may have alternative descriptors, for example, outline business case (OBC). The high-level strategic outline case is refined to reflect the benefits, costs and risks of the chosen option. At this stage of the project, it would be typical for estimates of benefits and costs to be a class 4 estimate (Guidance Note 14). The level of definition would be approximately 20% and with a typical accuracy of −30 to +50%. The client sponsor may engage specialist consultants, for example architects or commercial managers, to support the estimation process. Client organisations and funders, such as debt or equity investors and/or private investors, will have different requirements for the preparation of the business case, and practitioners must align their work with the client requirements.

I. **Document and approve stakeholder map.** Building from the high-level stakeholder analysis created in the Identify stage, the **client project manager** is responsible for producing a refined stakeholder map capturing known information about stakeholders (Guidance Note 6). The attitudes, influence and expected involvement in the project of individual stakeholders/ stakeholder groups is required, plus an understanding of the relationships between stakeholders and the social networks with the potential to influence stakeholder behaviour. The purpose is to be clear, before approval is given to move to the Develop stage, about individuals and groups who will have a potentially significant impact on the chosen option for the project (1) because they are influential and can impact decisions and progress or (2) because stakeholder views and behaviours can impact the realisation of benefits. This work is a collaborative activity, and the client project manager is responsible for capturing the views of representative **end users**, the **operator(s)**, funders, regulators and other influential internal and external groups.

Decisions

Before approving the project to move on from this stage, the relevant decision-makers as referred to in the governance or authority matrices are required to confirm:

1. The team has evaluated and recorded a wide enough range of sufficient options and used a logical and evidence-based process to select the chosen option based on information delivered by the team commissioned by the client sponsor. The selected options should be communicated to other key project stakeholders.

2. Key stakeholders and the relationships between stakeholders are understood and incorporated where necessary into clear, concise terms and conditions and arrangements.

3. The selected client options have the support of key stakeholders; they appropriately meet needs and are achievable in terms of funding and implementation. This includes visibility of the progress of planning consents. Agreements must be documented, agreed and signed off by named stakeholders.

4. The rationale for rejected options should be justified, recorded and communicated as necessary.

5. Known external dependencies, constraints and risks in the wider context are represented in the intermediate-level business case, logged and recorded in the logic-linked plan, the high-level risk register and risk management plan, with owners identified.

6. Deliverables adequately address each of the defined Themes as detailed below and should be recorded and distributed to key stakeholders.

    (a) **Quality** criteria and priorities are clear, agreed and recorded.

    (b) **Sustainability** design principles and performance measures are clear, agreed and recorded.

    (c) **Value** drivers and priorities are clear, agreed and recorded.

    (d) Assumptions regarding **productivity**, for example, the mix between on-site and off-site construction are clear, agreed and recorded.

    (e) The focus of **leadership** is on long-term principles and objectives, not short-term gain.

    (f) **Collaboration** with stakeholders and the supply chain is integral to the chosen options.

    (g) Information exists to validate the chosen option and information and **knowledge** sharing requirements are clear, agreed and recorded.

    (h) The **risks**, threats and opportunities are understood for the chosen option, and all parties understand the risk profile via a clearly defined and agreed risk register. The client's project manager is responsible for establishing this and managing the risks by engaging others in the project team as needed. (This continues through the follow on stages.)

The scope of work for the Define stage is understood and a suitably competent and experienced team is in place to do this work (Guidance Note 4) for example, consultants who can support procurement activities, experts in planning off-site manufacture/on-site assembly, specialist planners, risk analysts, structural engineers and mechanical, electrical and plumbing consultants. All team members should have clear and concise terms of reference and appointments or contracts of employment relative to roles and responsibilities.

3

Define: delivery approach and procurement strategy

The purpose of the **Define** stage of the life cycle is to plan the delivery of the chosen options to meet the defined needs and benefits. The procurement strategy is an integral part of achieving the delivery approach and should be developed in conjunction with key stakeholders, updated regularly and recorded via the PEP and in an accessible repository. As is always the case, a clear, concise master programme designating tasks, showing inputs and outputs and dependencies is fundamental to managing this stage effectively and efficiently. The guidance on Collaborative Procurement for Design and Construction to support Building Safety documents should be considered with regards to the procurement strategy.

The eight Themes are applied in the Define stage as follows:

- Incorporate **quality** criteria and priorities to plans and (any) construction tenders design enquiries ensuring clarity and the correct level of detail. At this stage, the stakeholders should now be evolving into a core project team.

- Incorporate **sustainability** design principles to plans and (any) tenders and contracts ensuring clarity and the correct level of detail.

- Re-clarify **value** drivers and priorities in the business case. Ensure the process of agreeing the value drivers has been recorded and agreed between the relevant stakeholders and the evolving core team whose roles and responsibilities will begin to be clarified in the PEP.

- Incorporate requirements for **productivity** and innovation to designs, and future tenders and contracts, for example expectations for Design for Manufacture and Assembly (DfMA). Detailed specification and off-site testing regimes must be fully recorded and monitored.

- Focus **leadership** to enable detailed roles, responsibilities, terms of reference and delegated limits of authority to be agreed and recorded in the PEP as part of the contractual and appointment evolution.

- Incorporate requirements for **collaboration** and risk allocation within appointments and future tenders and contracts and ensure these plans are accessible and updated at regular intervals.

3 Define

Code of Practice for Project Management for the Built Environment, Sixth Edition. Chartered Institute of Building.
© 2022 John Wiley & Sons Ltd. Published 2022 by John Wiley & Sons Ltd.

- Establish processes, such as administrative protocols, to gather and share **knowledge** and related information and learning with stakeholders (not just the immediate project team). These processes when validated in the PEP should be stored in an easily accessible repository.

- Final validation of **risk** strategy and plans, including use of financial and schedule contingency, technical redundancy, resilience planning and insurance requirements and ensure a risk register is monitored and updated in the PEP at regular intervals and stored in an easily accessible repository.

Activities

The purpose and application of the themes is achieved through implementation of the following activities and creation of documented deliverables.

The list of activities does not represent a strict sequence of work but is indicative of a likely order. In practice, many activities will happen in parallel and with iteration in order to create the deliverables needed to approve progression to the next stage of the life cycle.

A	• Establish the team and governance for the Define stage
B	• Define delivery approach
C	• Confirm procurement strategy (and any tender procedures)
D	• Document and approve project execution plan (PEP)
E	• Update and approve intermediate business case
F	• Document and approve change control process
G	• Document and approve stakeholder engagement and communication plans
H	• Procure consultants and contractors as required and confirm team for Design stage

A. **Establish the team and governance for the Define stage**. The **client sponsor** is responsible for commissioning a team who are suitably competent and experienced to complete the activities associated with detailed planning of the project. The team must include people/organisations with the appropriate information management capabilities and capacities, who have received a specific set of information requirements from the client (see Guidance Note 10). Specialists from consultancies or contracting organisations may be commissioned to support this work. In addition to core skills in design, surveying, procuring, construction, or project planning, experts with specialities in consenting, sustainability, information management or a particular regulatory requirement may be required. Note that information management in line with ISO 19650-2 is appointment oriented rather than project stage oriented. If one consultancy is appointed to assess the feasibility of options, their exchange information requirements would be limited to this scope, whereas another consultancy may have been involved in feasibility and on design, delivery and handover; therefore, their information requirements cover the whole scope. The **client project manager** is responsible for bringing the team together, taking responsibility for establishing effective working practices and leading the team to complete the activities in this stage to the satisfaction of the client sponsor and wider governance. It is critical all appointments contain easily understandable terms and conditions, clear and concise quantum matters and, most importantly, a defined scope of work.

B. **Define delivery approach**. The **client project manager** is responsible for agreeing the best methodology for the delivery of the chosen option. Initial decisions are focused on the organisational strategy, whether to produce/ deliver, buy or lease assets (delivery model assessments[1]), which is the conclusion of the validation of the business case. Where the client organisation is not involved, the procurement strategy needs to be established (see activity C). There are multiple options for organising work within the design, implementation, and validation stages of the project. A strict linear (waterfall) approach ensures the scope and quality of each preceding activity is fully delivered before starting the next. An iterative (agile) approach is where resources are organised into fixed time boxes and a fully dedicated team pursues as much scope to the right quality in the time-box before review by representative end users and iteration.[2] Some projects may be part of a wider master programme or portfolio within the client organisation and may share processes, practices and people with other projects.

Modern methods of construction, specifically Design for Manufacture and Assembly (DfMA), may enable more flexibility in defining the delivery approach than traditional methods, albeit the choice to manufacture off-site and assemble on-site may require a more stringent and defined approach.[3] Many projects may adopt a hybrid approach, for example, adopting a traditional linear approach to the design and build of the physical asset but with an agile approach to designing and building related digital capabilities by, for example, adopting the use of BIM or other relevant software platforms.

A key part of any delivery approach is the chosen strategy for procurement of goods and services, including the use of consultants and contractors. Works, goods and/or services may be packaged to ensure the reliance on any single supplier meets the risk appetite of the client organisation. Consideration should be given to the guidelines adopted by public sector organisations to ensure compliance with their governance protocols. Where such an approach is chosen, working practices may be put in place to incentivise collaboration and information sharing. Alignment between the delivery approach and procurement strategy is critical. Key design consultants must be appointed initially to ensure the design scope meets the contract requirements. Appointments should include clearly defined responsibilities for integration of designs in a fully coordinated manner, and this will ensure the procurement of contractors is based upon a robust design. It is important to be aware that off-site manufacturing processes and procedures are continuously improving and the project manager needs to be fully aware of the impact of innovation on time and cost, the value of off-site manufacture, and should consider the advantages of incorporating these processes into the works.

C. **Confirm procurement strategy (and any tender procedures)**. The **client project manager** is responsible for agreeing procurement strategies and any related tender procedures in conjunction with the definition of the delivery approach. An early contractor involvement approach could be advantageous

---

[1] Guidance Note 21: Delivery model assessment
[2] Guidance Note 22: Choice of project management approach/method
[3] Guidance Note 23: Impact of Design for Manufacture and Assembly (DfMA) on delivery approach

to ensure compliance in particular with regard to off-site modular construction. The adoption of such an approach would also have benefits with regards to promoting carbon reduction and sustainability requirements. Note the term procurement embraces all aspects of establishing and managing the supply chain for the project. The strategy will define the forms of contract to be used to source the external team and manage the apportionment of risk.[4] Decisions relative to tenders, frameworks and pre-qualifications should be made with due cognisance of the legal drafting process. These decisions must account for the capacity and capability in the market and/or any regulatory requirements or permissions to use local labour, local suppliers and to develop the skills of local people. A project requiring the support of the local community may need to guarantee jobs for local people or to create apprenticeships or other training to upskill local people for future jobs in the sector.

Improving procurement practices is a key strategic driver for the sector, moving beyond adversarial approaches when establishing and managing contracts. The procurement strategy must be designed to (1) help clients achieve needs and benefits efficiently and effectively, (2) share risk and make provision for contingency in the appropriate part of the supply chain, (3) incentivise innovation and (4) establish collaborative, long-term relationships for the benefit of future projects. In sharing risk and making provision for contingency, there is the need to ensure insurance cover is in place for the relevant supply chain member (see Guidance Note 26).

It is increasingly recognised procurement must establish a behavioural and cultural fit between the client and the key players in the supply chain. This particularly applies to the more technically complex and logistically challenging members (IT installations, complex works at height for example), in addition to the required technical and commercial competencies.[5] A high-level strategy for procurement may have been discussed ahead of this stage of the life cycle, where a framework agreement is already in place, where DfMA is integral to the project concept, or where a particular form of contract is the obvious choice. Where the procurement strategy includes the need to tender for work, for example in the public sector, the client project manager is responsible for managing the tender processes in line with all relevant legislation.[6] It is also best practice to agree the desired forms of dispute resolution at this stage.[7]

D. **Document and approve project execution plan (PEP).** The **client project manager** is responsible for documenting and approving the PEP[8], from inception. A major activity in this stage of the life cycle is the definition and implementation of plans to manage project scope, quality and compliance (including health, wellbeing and safety)[9], time[10,11], cost[12], risk[13,14,15] and

---

[4] Guidance Note 24: Forms of contract
[5] Guidance Note 25: Behavioural procurement
[6] Guidance Note 26: Tender procedures
[7] Guidance Note 27: Dispute resolution
[8] Guidance Note 28: Project Execution Plan indicative content
[9] Guidance Note 29: Scope and quality planning and management
[10] Guidance Note 30: Time planning and management
[11] Guidance Note 31: Resource planning and management
[12] Guidance Note 32: Cost/budget planning and management
[13] Guidance Note 33: Risk identification
[14] Guidance Note 34: Risk analysis and evaluation
[15] Guidance Note 35: Quantitative risk analysis and evaluation

contingency.[16] The PEP should include the plans to measure, monitor and control progress and performance[17,18], to manage change and dispute resolution[19] and to enable an efficient and effective construction process (see Guidance Note 28). It should also include protocols and processes for managing health and safety[20], procurement, risks and stakeholder communications. The PEP is one of the critical deliverables to be approved by the **client sponsor** and wider governance at the stage-gate review.[21] This triggers approval to move to the Design stage. After the Develop stage, the PEP is 'baselined' to set the reference point against which the project will be monitored and controlled via regular updates. Any changes must then be managed using change and configuration control procedures (see step F). It is a live document, typically available electronically to all relevant project participants and changes are managed by the project manager in accordance with pre-determined protocols. In effect, the PEP is the 'how' of how the team will apply the protocols to deliver the asset to meet the clients' requirements. It is imperative sufficient time and energy is devoted to the compilation and content of the PEP as compliance will be required by all stakeholders to the project team. In some environments, the PEP may be known as the Project Management Plan (PMP) or Consolidated Project Plan (CPP).

E. **Update and approve the intermediate business case.** The responsibility for this activity remains with the **client sponsor**; however, they may delegate the work to the **client project manager** to create the third iteration of the business case and agree this with the **client sponsor** and wider governance (Guidance Note 13). The term 'intermediate' business case is used again here, but this may have alternative descriptors depending on the client organisation, for example, 'outline business case'. The intermediate business case developed in the Assess stage is updated to reflect more detailed benefits, costs and risks of the chosen option. At this stage of the project, it would be typical for estimates of benefits and costs to be a class 3 estimate (see Guidance Note 14). The level of definition would be approximately 40% and with a typical accuracy of −20 to +30%. The client sponsor may engage specialist consultants, for example specialist architects or commercial managers to support the estimation process. Client organisations and funders will have different requirements for preparation of the business case and practitioners must align their work with the client requirements.

F. **Document and approve change control process.** The **client project manager** is responsible for documenting and approving the change control process[22] to maintain the PEP and the business case from this stage. It is itself a key part of the PEP. The versions of the PEP and Business Case at the end of the Define stage are considered to be the baseline, but they will be updated as more detailed work is completed in the Design stage using the change process already defined. Formal change control is required to ensure changes do not undermine any of the rationale for the project developed in the Identify and Assess stages and to maintain alignment between all

---

[16] Guidance Note 36: Contingency planning and management
[17] Guidance Note 37: Progress monitoring, measuring and reporting
[18] Guidance Note 38: Risk treatment
[19] Guidance Note 39: Issue resolution and problem solving
[20] Guidance Note 40: Health and safety plan.
[21] Guidance Note 41: Preparation for stage-gate reviews
[22] Guidance Note 42: Change control

parties involved in the project through their respective contracts and terms of reference. It is essential the change control process is recorded through a robust and agreed process, and all decisions and agreements are signed-off regularly between the parties. Failure to agree changes during the project life cycle will often lead to cumulative problems and behavioural issues that would have been avoided if good process had been followed. Identifying the roles and responsibilities for managing the change control process should be clarified in the PEP.

G. **Document and approve stakeholder engagement and communication plans.** Building from the stakeholder map created in the Assess stage, the **client project manager** is responsible for updating known information about stakeholders and expanding this into a plan for engagement and communication across the life cycle.[23] The communications plan is a key part of the PEP. This work remains a collaborative and dynamic activity, and the client project manager is responsible for capturing the views of representative **end users**, the **operator(s)**, funders, regulators and other influential internal and external groups noting this is a live document, and the external context for the project may change over time, for example new regulations, emerging technologies or changing societal expectations. All meetings related to stakeholder engagement should be recorded clearly and concisely, so that all parties are aware of the evolution of the scope of works and the appropriate responsibilities across the team.

H. **Procure consultants and contractors as required and confirm team for Design stage.** In line with the delivery and procurement strategy defined in this stage, the **client project manager** is responsible for ensuring there is agreement about the team for the Design stage defining roles, responsibilities and delegated limits of authority across the whole team, including consultants and contractors, and documenting these in the PEP. Several of these resources may have been involved in the Assess and Define stages of the project, for example architects and quantity surveyors, but formal confirmation of roles for Design is needed by governance before proceeding, for example, the inclusion of specialist mechanical and electrical engineers, or consultants with expertise in sustainable materials. Note that procurement must involve the preparation of information requirements (see Guidance Note 10) for each appointment so that delivery as required by the client has been resourced and costed. In confirming the team for the Design stage, the **client project manager** must ensure provisions are in place to ensure compliance with all relevant legislation, including but not limited to Health and Safety legislation within the jurisdiction in question.[24,25,26,27,28,29,30] It is recommended that standard unamended terms and conditions and appointments are used for all consultants and contractors. Amended terms and conditions may result in confusion, additional expenditure and potentially collateral damage to the parties involved. It is important the design team is procured

---

[23] Guidance Note 43: Stakeholder engagement and communications
[24] Guidance Note 44: Overview of the UK Health & Safety at Work etc Act 1974 (HSWA, 1974)
[25] Guidance Note 45: Overview of the UK Construction (Design and Management) Regulations 2015 (CDM, 2015)
[26] Guidance Note 46: Overview of the UK Dangerous Substances and Explosive Atmospheres Regulations, 2002 (DSEAR, 2002)
[27] Guidance Note 47: Overview of the UK Housing Acts
[28] Guidance Note 48: Overview of the UK Town and Country Planning Act (1990)
[29] Guidance Note 49: Overview of the UK Housing Construction and Regeneration Act 1996, Amended 2011
[30] Guidance Note 50: Overview of the UK Building Safety Bill (2021) and Fire Safety Act (2021)

as early as possible in the project life cycle to ensure a robust and accurate coordinated design for procurement purposes. The resource allocation would be part of the agreement prior to formal appointment. Consultants who will be procured from the identify to operate life cycle stages will be appointed for the total of their involvement, for example architects, mechanical and electrical consultants, landscape architects, BREEAM consultants, quantity surveyors, structural/civil engineers. Some specialists may be appointed for the early stages, or part stages only; planning consultants for example, may not be required at the latter stages of delivery and handover.

Decisions

Before approving the project to move on from this stage, the defined decision-makers on the governance board are required to sign-off all necessary milestones:

1. The delivery approach and procurement strategy is clear and evidence supports the use of unamended standard forms of contract related to the appointment of the supply chain to reflect market availability of resources and to generate the required levels of collaboration and risk sharing.

2. The delivery approach facilitates completion of the project in an agreed timescale, measured and monitored by a logic linked and accurate master programme.

3. The project execution plan (PEP) is sufficiently complete to form the baseline to move into Design activities, i.e. detailed planning of the Design phase and with sufficient definition of plans for subsequent stages to justify continuation, noting that the PEP cannot be updated until after detailed design activities are complete.

4. The updated intermediate business case, based on estimates in the PEP, justifies continued investment in the project.

5. Resources (contractors, sub-contractors, suppliers and the like) are available and costed to provide the structured and unstructured information relating to (1) the asset(s) and (2) the project to create the asset(s).

6. Detailed definition of process and controls for considering, justifying and documenting variations from the approved PEP and business case baseline are in place.

7. All stakeholder groups and key individuals are prioritised and methods of engagement and communication defined.

8. Deliverables adequately address each of the defined Themes.

   (a) **Quality** criteria and priorities are incorporated into plans and (any) tenders and contracts.

   (b) **Sustainability** design principles are incorporated into plans and (any) tenders and contracts.

   (c) **Value** drivers and priorities are incorporated into the business case.

   (d) Requirements for **productivity** and innovation are incorporated into plans and (any) tenders and contracts, for example expectations for Design for Manufacture and Assembly (DfMA).

(e) **Leadership** is in place, including the definition of detailed roles, responsibilities, terms of reference and delegated limits of authority as stated in the PEP.

(f) Requirements for **collaboration** and risk sharing in the supply chain are incorporated into plans and (any) tenders and contracts.

(g) Processes to gather and share **knowledge** and related information and learning with stakeholders and across the supply chain are established and recorded by the contractor and the supply chain.

(h) **Risk** strategy and plans, including use of financial contingency, technical redundancy, resilience planning and insurance requirements, are in place.

9. A suitably competent and experienced team is in place for the Design stage, including all relevant roles to comply with legislation, for example the Principal Designer in compliance with the current version of the Construction (Design and Management) Regulations[31] (see Guidance Note 51). As project outputs increase, then resources should be supplied to meet the design demand.

---

[31] Health and Safety Executive (2015), The Construction (Design and Management) Regulations 2015. Available at https://www.hse.gov.uk/construction/cdm/2015/index.htm (accessed 20 March 2021).

4 Design: specifications and functionality

Purpose

> The purpose of the **Design** stage of the life cycle is to determine the specifications and functionality of the asset in sufficient detail to enable manufacture and construction and in line with the delivery and procurement strategies. It is imperative design data between different design disciplines is fully coordinated to avoid interface and scope problems, and the use of appropriate technologies and processes such as BIM are adopted to ensure proper quality design.

The eight Themes are applied in the Design stage as follows:

- Translate key decisions and **quality** acceptance criteria into detailed specifications engaging relevant experts and stakeholders, document in the PEP and share with relevant parties.

- Confirm the design process and outputs uphold agreed **sustainability** commitments with relevant experts, document and share.

- Confirm the design process and outputs uphold the agreed **value** drivers and priorities in the business case with relevant experts, document and share.

- Translate the innovation strategy and **productivity** requirements into detailed specifications.

- Focus **leadership** on engaging stakeholders and resolving conflict according to agreed terms of reference and uphold health, safety, well-being and ethical standards in adherence to relevant legislation.

- Ensure design activities optimise the supply chain and balance **collaboration** with efficiency of delivery. Ensure decisions are documented and updated at regular intervals.

- Design in line with agreed **knowledge** and information management commitments and store design in an easily accessible repository.

- Implement **risk** plans through design activities, focus on and resolve emergent risks, keeping the risk register updated accordingly.

Code of Practice for Project Management for the Built Environment, Sixth Edition. Chartered Institute of Building.
© 2022 John Wiley & Sons Ltd. Published 2022 by John Wiley & Sons Ltd.

4 Design

Activities

The purpose and application of the themes is achieved through the implementation of the following activities and creation of documented deliverables.

The list of activities does not represent a strict sequence of work but is indicative of a likely order. In practice, many activities will happen in parallel and with iteration in order to create the deliverables needed to approve progression to the next stage of the life cycle.

A	• Establish the team and governance for the Design stage
B	• Design the asset(s)
C	• Assure and approve the design in line with defined governance
D	• Update business case, project execution plan (PEP) and other related plans
E	• Procure works, goods and services for implementation
F	• Procure (further) consultants and contractors as required for the Implementation stage

A. **Establish the team and governance for the Design stage**. The **client sponsor** is responsible for mobilising the team identified and approved in the previous stage (Chapter 3, Activity H). This team must comprise suitably competent and experienced people/organisations and subject matter experts to deliver the design activities associated with the needs and benefits of the client brief and in line with the delivery and procurement strategy. The team must have appropriate information management capabilities and capacities to deliver a specific set of information requirements from the client (see Guidance Note 10). Specialists from consultancies or contracting organisations may be commissioned to support this work. In addition to core skills in design, surveying, procurement, building, or project planning, experts with specialities such as building services, lifts, escalators and renewable materials may be required.

Note information management in line with ISO 19650-2 is appointment oriented rather than project stage oriented. If one consultancy was appointed to be involved in assessing feasibility of options, these exchange information requirements would be limited to this scope, whereas another consultancy may have worked on feasibility and is continuing to work on design and will be involved in handover, so these information requirements cover the whole scope. The **client project manager** is responsible for bringing the team together, taking responsibility for establishing effective working practices and leading the team to complete the activities in this stage to the satisfaction of the client sponsor and wider governance. This will include compliance with protocols specified in the PEP for stakeholder engagement and communication, monitoring, measuring and reporting progress and performance, and within-stage decision-making. Part of the client project manager's responsibilities include making it clear to the whole team and governance where the responsibility for design management resides – with an individual or an organisation, as part of, or separate from the design team. It is imperative the structure for governance and terms of reference for all relevant individuals are agreed, recorded and stored in an easily accessible repository. Governance levels and limits of authority should be regularly reviewed to reflect the progress and volume of work being carried out.

B. **Design the asset(s).** The **client project manager** is responsible for leading the design team through a clearly established design process, including iterated verification and validation of all functionality and production of all information identified in the designers' information delivery plans, for example detailed design models and drawings incorporating detailed structural, energy and acoustic analysis, in addition to schedules of components. They are also responsible for liaising on the above with the end user/operator. Good design is fundamental to sustainable development and plays a crucial role in how places are perceived, with the quality of the built environment contributing to a positive perception from stakeholders. Decisions made about design will impact safety, the environment, ecology, customer experience, the economy and wider society, for example the materials used for construction, the layout of a communal space or the aesthetic of the design in promoting tourism. Good design builds on fundamental principles[1], defines the design in increasing levels of detail[2] and ensures the design process is suitably controlled. Use of coordinated design processes and technologies will help to ensure this happens. The design management process relies on good quality information on which to base assumptions. One critical area is the information produced from site investigations (see Guidance Note 19). The project manager should take guidance from all available experts to validate a competent design process but should always be aware of the innovative and ever-changing methods of construction. Therefore, advice will need to be continually reviewed and updated to reflect new methods and technologies.

C. **Assure and approve the design in line with defined governance.** The **client project manager** is responsible for commissioning assurance to verify and validate the completeness and technical achievability of the design, and to ensure the needs and benefits of the chosen options may be achieved through design. Assurance activities must satisfy the needs of governance and the three lines model (see Guidance Note 3) and implement the designated quality plans within the PEP (see Guidance Note 28). The audit trail of design work and decisions throughout the implementation of the design process and contained in design information will form the basis for assurance, including, but not limited to, material specifications, drawings, calculations, meeting notes and so on. In some circumstances, additional assurance will be provided through peer review by a competent authority and/or regulatory approvals, for example the planning authority, or Building Control. In many projects, and depending on the form of contract, responsibility for the management and delivery of design activities will be undertaken by consultants, or contractors, for example a 'design and build' type contract will empower the contractor. Where this is the case, the client is sufficiently independent from the work to provide technical assurance should they have the technical proficiency to do so. In all other situations, the client should procure further independent consultants to provide assurance on its behalf.

---

[1]   Guidance Note 51: Design management fundamentals
[2]   Guidance Note 52: Detailed design

4 Design

D. **Update business case, project execution plan (PEP) and other related plans.** The responsibility for the business case remains with the **client sponsor**; however, they may delegate the work to the **client project manager** to create the final iterations of the business case before the investment decision, which triggers the majority of the spend in Implementation (see Guidance Note 13). As before, the business case must be agreed with the **client sponsor** and wider governance. Updates to the business case should be reflected in the PEP and other related plans, under formal change control, ensuring the PEP and business case remain aligned. In some organisations, the business case and PEP may be combined, but this is not typical given the business case is client confidential, whereas the PEP is a document used in the mobilisation and management of the wider client/consultant/contractor team. The level of definition at the end of the Design stage should have a typical accuracy of −10 to +15% (class 1 estimate) (see Guidance Note 14), with the financial contingency defined and approved (see Guidance Note 36). It is common, depending on the project, for an interim update to be prepared resulting in a class 2 estimate at approximately 70% definition, and with a typical accuracy of −15 to +20%. In some cases, multiple milestones and decision-points may be set within the Design stage to progressively assure the design and improve the accuracy of estimates in the business case. This is because the decisions made at the end of the design stage-gate are usually the final investment decision for the client, after which substantial abortive costs would be applicable should the design prove not fit for purpose or the costs, risks and benefits of the business case become unrealistic.

E. **Procure works, goods and services for Implementation.** Once approved, the **client project manager** is responsible for managing the procurement of all goods, materials and services required for the Implementation phase (see Guidance Notes 25–27). In many projects, depending on the form of contract (Guidance Note 24), it is more often the contractor who will take responsibility for the procurement of a significant sub-set of the goods and/or services in accordance with the specific information requirements issued to each supplier (see Guidance Note 10). Depending on the form of contract, the risks associated with the cost of purchases are likely to reside with the contractor; however, the client project manager retains overall responsibility for ensuring goods and services procured meet all relevant quality and compliance requirements and are provided in a timely manner to facilitate the achievement of the agree project time-line. Prior to final approval of the business case, approval may be granted to the client project manager to sanction early procurement of scarce resources and long lead-time items required for implementation. It is imperative that procurement adheres to the agreed procurement strategy, that robust enquiry documentation is sent to suppliers in good time and is managed by logic-linked and accurate procurement programmes.

F. **Procure (further) consultants and contractors as required for the Implementation stage.** The **client project manager** is responsible for implementing the procurement strategy (Guidance Note 25) and ensuring the team required for the Implementation stage (as documented in the budget contained in the PEP (Guidance Note 28) and business case (Guidance Note 13)) is resourced through the internal client, consultants, contractors and their sub-contractors, with input from stakeholders as required. Before leaving the Design stage, the client sponsor and wider governance require assurance from the client project manager that suitably competent and experienced resources are in place to manage the Implementation phase. This is a pivotal point in the project, where the major investment decision is taken and commitments are made to key stakeholders about the quality of delivery and the timescales to the provision of the asset(s). The size of the overall project team will increase significantly from this point and will include those whose labour will be involved in the manufacture, assembly and construction of the asset. It is important the client project manager communicates with all team members across the supply chain to ensure they are aware of the plans and controls embedded in the integrated PEP, not only their particular contract.

Decisions

Before approving the project to move on from this stage, the relevant decision-makers are required to confirm:

1. The asset(s) as designed will meet the needs and benefits and in accordance with all relevant standards and regulations as defined in the quality plan.

2. The nature and size of variations that will be tolerated and whether these are within the appetite for risk and approved contingencies, or whether changes to risk appetite, and any associated financial contingencies need to be approved by corporate governance. Variations to all the intended needs and benefits require approval, not only the financial aspects.

3. The design information produced is correct, complete and consistent, in accordance with the relevant information requirements issued by the client.

4. The project execution plan (PEP) is up to date and in sufficient detail to guide successful implementation.

5. The business case remains viable.

6. Stakeholders are engaged, any issues are being resolved and do not threaten the continuation of the project recognising that in the transition to the physical build some stakeholders, such as the planning authorities, become less critical and others, such as Building Control, become more so.

7. Deliverables adequately address each of the defined Themes.

   (a) Key decisions and **quality** acceptance criteria are clear in detailed specifications.

   (b) Design upholds agreed **sustainability** commitments.

   (c) Design upholds the agreed **value** drivers and priorities in the business case.

   (d) Innovation strategy and **productivity** requirements are clear in detailed specifications.

   (e) **Leadership** is focused on engaging stakeholders, conflict resolution and upholding health, safety, well-being and ethical standards.

   (f) The design optimises the effectiveness of the supply chain, balancing **collaboration** with efficiency of delivery.

(g) The design upholds agreed **knowledge** and information management commitments.

(h) The design resolves as many identified significant issues and **risks** as practicable, for example fire risks posed by materials choices.

8. A competent and experienced team is in place for the Implementation stage, including all relevant roles in compliance with legislation, for example the principal contractor as defined within the current Construction (Design and Management) Regulations (see Guidance Note 45). It is imperative terms of reference and appointments provide a clear and concise description of the roles and responsibilities.

5 Implement: manufacture and construction

Purpose

> The purpose of the **Implement** stage of the life cycle is to manufacture and construct the asset in compliance with the design and delivery and procurement strategies, plans and associated contracts. As discussed in previous stages, the implementation should be carried out to a robust plan that has been reality tested and logic linked.

The eight Themes are applied in the Implement stage as follows. It is recommended that information about their application is documented and shared with known project stakeholders, in an easily accessible repository.

- Deliver **quality** by manufacturing and constructing to specification. Continually monitor against documented specifications, update where agreed adjustments are made.

- Uphold agreed **sustainability** commitments through manufacturing and construction and ensure they adhere to client requirements.

- Track performance and implement change control to justify the **value and impact** of any changes. Ensure the process and agreed change have been recorded and agreed between the relevant and appropriate stakeholders.

- Deliver innovation and **productivity** targets through manufacturing and construction. Ensure these meet or exceed time, quality and value requirements.

- Focus **leadership** on engaging stakeholders and conflict resolution according to agreed terms of reference and upholding health, safety, well-being and ethical standards in adherence to relevant legislation.

- **Collaborate** across the supply chain to ensure quality of implementation ensuring activities with the potential to delay or impact others are regularly and clearly communicated across the teams or supply chain.

- Manufacture and construct in line with agreed **knowledge** and information management commitments. Monitor compliance, update data and information and update the PEP.

- Implement **risk management** procedures throughout manufacturing and construction activities, resolve issues, update and maintain a live risk register.

Code of Practice for Project Management for the Built Environment, Sixth Edition. Chartered Institute of Building.
© 2022 John Wiley & Sons Ltd. Published 2022 by John Wiley & Sons Ltd.

Activities

The purpose and application of the themes is achieved through the implementation of the following activities and creation of documented deliverables.

The list of activities does not represent a strict sequence of work but is indicative of a likely order. In practice, many activities will happen in parallel and with iteration to create the deliverables needed to approve progression to the next stage of the life cycle (Validate). There should be a continual awareness of innovation and technology improvements, and ongoing research should be carried out to ensure that best current practices are adopted.

A	• Establish the team and governance for the Implementation stage
B	• Manufacture and construct asset(s) safely
C	• Measure, monitor and manage performance
D	• Assure and approve as-built asset including any changes in line with governance
E	• Document operations and maintenance plans and manuals
F	• Procure works, goods and services for Operation and Maintenance
G	• Procure (further) consultants and contractors as required for the Validate stage

A. **Establish the team and governance for the Implementation stage**. The **client's project manager** has a duty to make sure the client has fulfilled all their duties before letting construction activity commence. The Principal Designer should be part of the corporate governance process. The **client sponsor** is responsible for commissioning a team who are suitably competent and experienced to manufacture and construct the design in line with the delivery and procurement strategies, plans and associated contracts. The team must include people/organisations with the appropriate information management capabilities and capacities who have received a specific set of information requirements from the client (see Guidance Note 10). Note that information management in line with ISO 19650-2 is appointment oriented rather than project stage oriented. Any person or organisation joining the project at this stage must be issued with their specific exchange information requirements by the client. This is the stage of the project life cycle when the team is substantial with a large number of specialist consultancies and contracting organisations involved in the provision of the works. The **client project manager** is responsible for finalising all contracts (see Guidance Note 24–26), bringing the team together, taking responsibility for establishing effective working practices and leading the team to complete the activities in this stage to the satisfaction of the client sponsor and wider governance and in line with all relevant legislation. Whilst some contracts may have started ahead of this stage, the client project manager must focus on building and/or maintaining supply chain relationships, ensuring positive outcomes for all contractual relationships. Beyond individual contracts, the **client project manager** must keep focused on interdependent elements of the overall system and the integration of work performed by multiple parties, anticipating the need to validate working systems in the Validate stage of the life cycle. The **client project manager** must ensure compliance with protocols specified in the project execution plan (PEP) taking specific responsibility for facilitating stakeholder engagement and communication (Guidance Note 43), monitoring, measuring and reporting progress and performance against the approved master programme (Guidance Note 37), and within-stage decision-making.

B. **Manufacture and construct asset(s) safely**. The **client project manager** is responsible for the management of the parties contracted to manufacture and construct the asset(s), including any refurbishment or demolition work in scope for the project. The **client project manager** is also responsible for providing leadership in the areas of health, safety and well-being. They must pay close attention to near misses (for example a fall not resulting in any injury) and to reportable or non-reportable incidents such as lost time, accidents, leakages into waterways or diseases such as legionella. Safety specific roles are required to ensure compliance with current legislation, the Principal Contractor for example, as defined by the Construction (Design and Management) Regulations[1] (see Guidance Note 45). Upholding the agreed sustainability commitment, waste management, noise pollution and/or general public relations regarding the impact of the works on the wider community, for example, are priorities. In many projects, the works will involve a combination of off-site manufacture and on-site assembly and construction. Projects will vary, but this is further discussed in Guidance Note 53.[2] Examples of specific information requirements to support an on-site/off-site implementation methodology might include the size and weight of off-site manufactured elements and the lifting-eye locations to enable cranage to be planned. It is important logistics management and buildability is fully addressed. The client's project manager should set the competency regime by setting up and implementing protocols to check and approve all proposed contractors and sub-contractors based on evidence of their competency, and likewise, evidence of accreditations for some of the service providers and proposed materials. For the long-term safety of occupants in the built asset, quality and safety management go hand in hand, and for example the setting up of the requirements for testing and commissioning in a safe manner.

C. **Measure, monitor and manage performance**. The **client project manager** is responsible for ensuring performance is measured, monitored and reported at multiple levels, for example from the main contractor to the client, or the client project manager to the client governance board. Actual and planned performance should be monitored and the benefits, costs and risks agreed in the business case should be tracked. In addition, the client project manager should monitor performance in accordance with individual contracts in the supply chain. The progress of the physical build and associated information should be monitored in accordance with the approved design by the client project manager. Subject matter experts employed as part of the project team should be deployed to monitor various aspects of the work, for example project controls experts, quantity surveyors, clerks of works and other inspectors. For different types of construction projects, it is important to define the roles and responsibilities of the specialist consultants, for example a mechanical, electrical and plumbing consultant may have more input to a commercial office construction than a structural refurbishment

---

[1] Health and Safety Executive (2015) The Construction (Design and Management) Regulations 2015. Available at https://www.hse.gov.uk/construction/cdm/2015/index.htm (accessed 20 March 2021).

[2] Guidance Note 53: Off-site/on-site considerations

scheme. It is increasingly the case that digital methods of monitoring progress will be employed on-site, for example drones and potentially, in the future, robotic technology. Additionally, some technology-driven construction methods may enhance real-time performance, such as 3D product design and robotic material installation. Despite advances in technology, there is also a vital role to be played by 'walking tours' of the site, not only to provide visible safety leadership but also for picking up defects. Performance found to be below expectation of quality and contractual agreement must be resolved via dialogue or resolution protocols. Where a dispute arises, alternative dispute resolution methods are preferable to litigation, and the **client project manager** is responsible for implementing an approach to preserve quality and value for the project, as well as ensuring trust and fair relationships within the supply chain (see Guidance Note 29). This is important where a project may be reliant on a contractor or consultant because of reduced competition in the market, or where a supplier is important to the client's wider portfolio and business interests. A robust data management system should be adopted and correct and timely records provided to accurately document the actual progress of the works.

D. **Assure and approve as-built asset including any changes (and associated contract variations) in line with governance.** The **client project manager** is responsible for working with the overall team to ensure the quality of the as-built asset complies with the design intent and upholds all elements of social, environment and economic value specified. Where change is required, or where non-compliance has occurred that need to be considered as concessions, or re-work, the client project manager is responsible for ensuring the change control process is upheld, and there is a clear audit trail of information (see Guidance Note 42). The overall assurance of the as-built asset requires the assurance of the as-built information. The information should be kept up to date and include any changes made to the asset during construction. This is a key role for the main contractor as a lead appointed party under ISO 19650-2 (see Guidance Note 10). As-built information includes models, drawings and documentation that records what has been constructed and installed, and any agreed changes made during this stage. Data storage and record keeping should be a key consideration.

E. **Document operations and maintenance plans and manuals.** In readiness for handover in the Validate stages of the life cycle, the **client project manager** is responsible for ensuring operations and maintenance (O&M) plans and manuals[3] are in place. The O&M manual is an authoritative guide to the asset as built and should include design principles, materials used, as-built drawings and specifications, asset registers of plant and equipment, operations and maintenance plans[4] and instructions, user guidance, health and safety information, testing results, guarantees, warranties and certificates, assumptions on whole-life costs[5], requirements for demolition, decommissioning and/or disposal. The O&M manual is the starting point for information about the asset at the commencement of operational use. In addition to the O&M manual, it is good practice to provide a building owner's manual or a user

---

[3] Guidance Note 54: Operations and Maintenance (O&M) manual

[4] See British Standard 8210: 2020 'Facilities maintenance management. Code of practice' for information about setting up maintenance strategies.

[5] See British Standard 8544: 2013 'Guidance for life cycle costing of maintenance works'

guide (or equivalent for non-building assets) to communicate to the Operator the purpose for the asset and its operational use. The knowledge of the design and build team should allow the operator to manage the building as safely and efficiently as possible, for example, to optimise energy usage.

F. **Procure works, goods and services for Operation and Maintenance.** This role could be commenced via a liaison between the clients' project sponsor and the end users' facilities operator. Eventually, the role will be carried out by one of the above individuals. Once Operations and Maintenance plans are approved, the **client project manager** or **facilities operator** is responsible for managing the procurement of all works, goods, materials and services required for the Operate stage of the life cycle, for example an outsourced facilities management (FM) provider. Depending on the form of contract (see Guidance Note 24), contractors may take responsibility for the procurement of a sub-set of the goods and/or services in accordance with the specific information requirements issued to each supplier, for example spare parts. The risks associated with the cost of purchases are likely to reside with the contractor; however, the client project manager retains overall responsibility for ensuring goods, and services procured meet all relevant quality and compliance requirements and are provided in a timely manner to support the transition of the asset into operation. For some projects, operation of the asset(s) may not be the responsibility of the client and the client focus will be to prepare the asset(s) for sale. Procurement for operation and maintenance works should be managed by the use of a robust logic-linked and accurate procurement programme.

G. **Procure (further) consultants and contractors as required for the Validate stage.** Consideration should be given to the appointment of specialist consultants early to avoid delays and problems in the latter part of the project. The **client project manager** is responsible for the delivery of the procurement strategy and ensuring the team required for the Validate stage (and budgeted for in the *business case* and *project execution plan*) is resourced through suitably competent and experienced internal client staff, consultants, contractors and their sub-contractors. Examples of additional team members required for the Validate stage are specialist engineers required to commission particular systems, such as ventilation. Stakeholders more specifically engaged in the Validation, and other stages may be, for example, building regulations, building control, legislative bodies and statutory undertakers. Although involved throughout all project stages, their role in approving the final deliverable is critical, and these resources should be co-ordinated to enable the timely transition of the asset from construction into operation. The size of the overall project team will decrease at this stage although many contractors and consultants will continue to have liabilities beyond the completion of the physical works and to the conclusion of defined defects periods. It is important the **client project manager** continues to collaborate and provide leadership to the team in concluding their contractual obligations.

Decisions

Before approving the project to move on from this stage, the relevant decision-makers are required to confirm:

1. The physical asset(s) as built is complete, and complies with the design, all relevant standards and regulations and all other obligations defined in the quality plan.

5 Implement

2. The asset information is correct, complete and consistent, in accordance with the client information requirements.

3. Any changes have been formally approved, and the impact of these on the intended needs and benefits understood. Changes that breach the defined risk appetite and approved contingency have been recorded, reported and approved by corporate governance.

4. Stakeholders are engaged to enable the move into Validation and Operation.

5. Deliverables adequately address each of the defined Themes.

   (a) Manufacturing and construction **quality** is to specification.

   (b) **Sustainability** commitments have been measured and upheld.

   (c) The **value** of any changes can be justified.

   (d) **Productivity** targets were measured and upheld through off-site/on-site working.

   (e) **Leadership** has engaged stakeholders, resolved conflicts and upheld health, safety, well-being and ethical standards.

   (f) **Collaboration** across the supply chain ensured efficiency and quality of delivery.

   (g) Information management commitments have been upheld and **knowledge** enhanced through demonstrable learning.

   (h) **Risks** associated with the next stages in the life cycle are clearly identified and assessed, with suitable risk treatments in hand.

6. A suitably competent and experienced team is in place for the Validate stage, for example independent consultants to validate compliance of systems.

Validate: integrate and handover

Purpose

The purpose of the **Validate** stage of the life cycle is to finally validate the integration of all systems, confirm the specification and functionality of the integrated asset through commissioning activities and to hand over ownership of the asset to the operator. It is critical the correct resources are allocated to the validation stage as failure to do this can result in additional cost and unrecoverable time. This stage should be adopted for complex projects and adapted as the client sponsor sees fit for less complex projects.

The eight Themes are applied in the Validate stage as follows:

- Validate the **quality** of the as-built asset in terms of specification and all associated regulatory requirements. Identify any non-compliance and agree resolutions with relevant parties. Update and share plans with key stakeholders and update relevant handover documentation.

- Validate that agreed **sustainability** commitments have been met through the build and plans for ongoing operation and maintenance. Identify any non-compliance and agree resolutions with relevant parties. Update and share plans and relevant handover documentation.

- Establish the benefit measurement baseline and ongoing tracking of performance to enable the planned **value** to be demonstrated over time. In essence, performance is compared with value. Ensure inclusion in handover documentation.

- Ensure innovation and **productivity** requirements for operations are signed-off and achievable. Ensure inclusion in handover documentation.

- Advise operational **leadership** is set up for success before they take ownership of the asset.

- **Collaborate** with stakeholders to ensure all key voices (clients, operators, end-users and occupiers) are heard during integration and handover activities through detailed recording of agreed actions and updated handover documentation.

- Capture **knowledge** and learning from manufacture and construction to improve contents of health and safety files and supporting operations and maintenance plans. Validate information deliverables are to specification and share with all interested parties (clients and operators).

Code of Practice for Project Management for the Built Environment, Sixth Edition. Chartered Institute of Building.
© 2022 John Wiley & Sons Ltd. Published 2022 by John Wiley & Sons Ltd.

- Implement **risk** plans through integration, validation and handover activities resolving issues and keeping the risk process alive. Ensure risk register is updated with final all actions taken and then handed over to operational management for ongoing ownership.

Activities

The purpose and application of the themes is achieved through the implementation of the following activities and creation of documented deliverables.

The list of activities does not represent a strict sequence of work but is indicative of a likely order. In practice, many activities will happen in parallel and with iteration in order to create the deliverables needed to approve progression to the next stage of the life cycle.

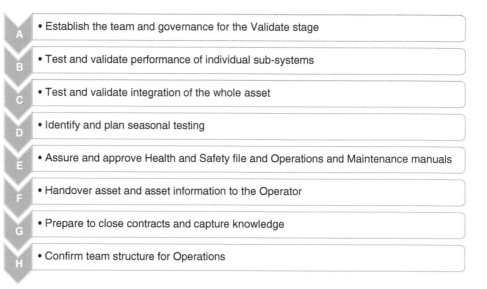

A   • Establish the team and governance for the Validate stage

B   • Test and validate performance of individual sub-systems

C   • Test and validate integration of the whole asset

D   • Identify and plan seasonal testing

E   • Assure and approve Health and Safety file and Operations and Maintenance manuals

F   • Handover asset and asset information to the Operator

G   • Prepare to close contracts and capture knowledge

H   • Confirm team structure for Operations

A. **Establish the team and governance for the Validate stage.** The **client sponsor** is responsible for commissioning the team who are suitably competent and experienced to validate, integrate and handover the built assets, ensuring the operator has all they need to manage ongoing use of the asset to deliver the needs and benefits identified. Note the operator may be an in-house estates team, an outsourced facilities management company or a third-party post sale or lease. The team must include people/organisations with the appropriate information management capabilities and capacities who have received a specific set of information requirements from the client. Note that information management in line with ISO 19650-2 is appointment oriented rather than project stage oriented (see Guidance Note 10). Any person or organisation joining the project at this stage must be issued with relevant and specific exchange information by the client. Team size will reduce throughout this life cycle stage; however, the client will be reliant upon specialist consultancies and contracting organisations to support the work, several of whom may be continuing from earlier life cycle stages, for example architects, mechanical and electrical engineers, structural engineers and builders of the asset. Other specialists may join at this stage, for example independent consultants to validate compliance of systems with the specification including regulatory authorities. Legal experts should also be involved to approve that as-built assets are satisfactory to prepare for the smooth transition to the operator in line with the Soft Landings philosophy (see Guidance Note 9). Members of the future operations team should be involved to set up the maintenance regime or, for example, to oversee and coordinate work associated with physical migration management whenever this occurs. The **client project manager** is responsible for finalising completion contracts, bringing the team together, taking responsibility for establishing effective working practices and leading the team to complete the activities in this stage to the satisfaction of the client sponsor and wider governance and in line with all relevant legislation. This stage of the life cycle requires the co-ordination and integration of sub-systems, which typically requires different suppliers to work together collaboratively. The **client project manager** must ensure compliance with protocols specified in the *project execution plan* (PEP) and associated contracts, with specific responsibility for facilitating stakeholder engagement and communication (see Guidance Note 43) monitoring, measuring and reporting progress and performance (see Guidance Note 37) and within-stage decision-making.

B. **Test and validate performance of individual sub-systems.** The **client project manager** is responsible for ensuring the work as specified and planned is implemented and documented in the *project execution plan*. For many projects, testing and validation may be phased over time, and thus some elements become operational ahead of others, for example where phased occupancy is a requirement or where an existing asset is being refurbished or extended. The project manager must ensure no party is prevented or hindered from commencing, continuing or completing their contractual obligations, and there is no unplanned interference with the effective operation of any existing assets. Procedures for testing and validating will vary depending on the nature of the asset; however, they should always define and provide data on compliance and non-compliance (for example unfinished work, weather damage or sub-standard performance). Increasingly individual sub-systems will include digital capability, which must be demonstrated and documented in information deliverables. Gantry signals on smart motorways, handling equipment in a warehouse or application of the 'internet of things' in housing such as smart heating or video doorbells, for example, would be documented in the project specifications. Where remedial work is needed, the client project manager is responsible for ensuring this is carried out in accordance with the terms of existing contracts (Guidance Note 24) or by virtue of change control procedures. Insurance claims may also be initiated, if appropriate.

C. **Test and validate integration of the whole asset.** The **client project manager** is responsible for ensuring the work is planned (documented in the *project execution plan*) and implemented. When individual sub-systems have already been validated (following re-work/remediation as necessary), system integration can be validated and regulatory approvals granted. Examples might be Environmental Health approval of a kitchen facility, or a Fire Safety approval of escape routes. The greater the number of interdependent systems, the greater the challenge in integration. Some projects may require specific resources assigned to focus upon the collective performance of separate elements, for example a large railway scheme organises their integration approach to meet four principles — collectively safe, operable, maintainable and performing. Where a project has been refurbished or has extended existing assets, safe and effective integration and integration

6 Validate

testing of new and existing systems are fundamental. The concept of commissioning, the bringing of something newly produced or modified into working condition, is frequently used to describe the activities required to validate that the asset meets and will deliver the intended benefits. Testing covers the methodologies that can be utilised to validate fitness for purpose, for example testing of equipment or feedback from users. Example commissioning checklists are included in guidance notes from a client handover perspective[1] (Guidance Note 55) and a client commissioning checklist from a building services perspective is included as Guidance note 56[2] noting the detail of such a document is dependent on the nature of the asset.

D. **Identify and plan seasonal testing**. The **client project manager** is responsible for working with the **operator** to identify those systems which need to be commissioned in season, for example heating systems, air-conditioning systems, motorised natural ventilation systems. The objective is to prove the capability of the systems at the extremes of the design envelope. Appropriate simulation of the change in season may be possible, for example adding heat loads to simulate summer conditions. Testing may also be planned for a later date, when actual conditions apply. Testing seasonal systems may be done by retained members of the project team or by an independently contracted consultant. Post-handover responsibility remains with the client to ensure all systems are fit for purpose, appropriately used and operating effectively across the seasons. It is critical when programming the construction and procurement of the works that due cognisance is given to the seasonal testing periods.

E. **Assure and approve Health and Safety file and Operations and Maintenance manuals**. The **client project manager** is responsible for ensuring, prior to the conclusion of any design and build related contracts, the health & safety file and any operations and maintenance (O&M) manuals are in place and reflect the actual installation and performance of each individual system of the overall as-built asset. The O&M manuals will have been in development from the design stage of the life cycle, and in this stage, these are finalised. This must include updates as necessary to ensure the O&M manuals describe the actual performance of the various elements of the asset as tested (see also activity F). The O&M manuals must include, as a minimum, design principles, materials used, as-built drawings and specifications, asset registers of plant and equipment, operations and maintenance plans and instructions, user guidance, health and safety information, testing results, guarantees, warranties and certificates, assumptions of whole-life costs, requirements for demolition, decommissioning and/or disposal (see Guidance Note 54). The Construction (Design and Management) Regulations 2015 (see Guidance Note 45) stipulate the Principal Designer is responsible for the preparation/ provision of the Health and Safety File, which refer, where necessary, to the relevant O&M manuals to ensure the safety of any party working on the asset, for example cleaning and maintaining the roof, or altering, refurbishing, extending or renewing the asset, in future. They must then ensure it is appropriately reviewed, updated and revised to take account of the construction works and any changes that have occurred. The Health & Safety File with any relevant O&M manuals are the starting point for information about the

6 Validate

---

[1] Guidance Note 55: Client handover checklist – indicative content
[2] Guidance Note 56: Client commissioning checklist – building services example

asset across the operational stage. Should the client project manager receive information from the client and/or contracted parties advising there are more efficient health and safety and operational and maintenance procedures, these should be considered and incorporated into updated Health & Safety and supporting O&M manuals.

F. **Handover asset and asset information to the Operator**. The **client project manager** is responsible for organising the handover of the asset and all related information to the operator when the asset is ready for use. The project manager must ensure the operator is aware of what assets are installed and the performance criteria and must include provision of a working performance management system, including baseline data. Training is clearly a key part of handover to ensure the operator has confidence in the use of the asset to realise the intended benefits. Contracts (see Guidance Note 24) must include a stipulated period of time for rectification of any defects. As per the soft landing philosophy (see Guidance Note 9), there may be a stipulated period of extended aftercare to achieve the intended operating performance and benefits. In this case, the asset performance management system and baseline data would be handed over at the end of the extended aftercare period. It is however critical any extensions to the aftercare period are agreed, including all cost and time implications, in advance of commencement. Phased handover may have significant advantages for the operator; however, contracts must reflect the need to provide support in partnership with operators. The client may commission a facilities management company[3] to provide a service to manage the ongoing asset operation and maintenance. Facilities management encompasses the traditional estate management functions including property maintenance, lighting, heating, security, fire safety, catering, logistics, for example. There is increased focus on spatial analytics, the planning and review of occupation and activity flows throughout the asset. Facilities may include all the built assets, furnishings, equipment and the environment available to end users who will occupy and use the asset.

Asset management is the term more commonly used for infrastructure. Facilities and asset management services are greatly enhanced by digital technologies that allow performance data to be continually captured, stored and analysed. It is increasingly the case that computer-aided facilities management software may be used to plan and track assets and maintenance, and to interface with building management systems reliant on sensors to provide feedback on asset performance. The use of big data enables the proactive identification of trends to anticipate future potential operational needs. Where the asset is to be sold, the same rigour is needed to achieve a soft landing; however, this needs to be integrated with the relevant operational activities to make the sale and ensure the acquiring organisation has the relevant data to realise the benefits for end users.

G. **Prepare to close contracts and capture knowledge.** The **client project manager** is responsible for administrative closure of contracts further to the physical completion of works. Most contracts will be nearing conclusion in the Validate stage. Some contracts may cease here, for example

---

[3] See British Standard 8572: 2018 'Procurement of facility-related services. Code of practice' to support decisions about facilities management.

hire of temporary accommodation; however, it is usual for the main contracts supporting design and build to include a defects liability period requiring an interface with the operational use. These contracts will continue to be managed by the client project manager in order to close out contractual obligations and defective works. It is critical to recognise where any extension of time may be required, subject to contract provision. Where disputes arise within the supply chain, the **client project manager** is responsible for managing these through to conclusion (see Guidance Note 27). Capturing knowledge and learning lessons from project activity is a guiding principle to be adopted across all life cycle stages, documented and shared. The client project manager is responsible for ensuring lessons are captured and actions are taken to change practices and processes to ensure lessons learnt are applied in future projects. At the end of the project, the principal designer, or where there is no principal designer, the principal contractor, must pass the health and safety file to the client.

H. **Confirm team structure for Operations.** The **client** or subsequent owner must advise the operational management team what resources are required to operate, to support operations and to maintain the asset as described in the *O&M Manual*. Resources may be outsourced, for example to a facilities management company or series of contractors, or may be client employees, direct labour, either reporting directly to the operations manager, and/or provided through an in-house asset management centre of excellence. Where existing project resources are needed to support early operations, for example fine-tuning systems, the **operator** must ensure there is a clear understanding of requirements, how contracts will be managed through to successful completion and close out, and how payments, liabilities and insurances will be managed. Service-level agreements and performance-based appointments ensure the transition to the Operate stage is fit for purpose. It is important the agreed terms of reference and scope of works are defined, written and agreed for the operational team structure.

Decisions

Before approving the project to move on from this stage, the client or subsequent owner is required to confirm:

1. The asset(s) as built and all associated information (including asset lists, building information models) perform as designed, meet the needs and benefit the client, operator and end users. Where seasonal commissioning or post occupancy evaluation is required, this is clearly identified.

2. Where the client will operate the assets, any changes from the design and the impact on the intended needs and benefits are understood. Changes or non-compliance that breach the defined appetite for risk and approved contingency have been reported to corporate governance and are approved. The asset may be sold post-handover; however, the work to prepare assets for sale is out of the scope of this *Code of Practice*.

3. Stakeholders are engaged across the supply chain, and any matters such as defects, guarantees or warranty issues for the end users are being resolved and do not threaten successful operation of the asset.

4. Deliverables adequately address each of the defined Themes.

   (a) **Quality** is validated – to specification and in line with regulatory requirements.

   (b) **Sustainability** commitments are met and continued into Operations and Maintenance Plans.

   (c) Baseline data and a system is in place to measure the **benefit** of the asset in use.

   (d) Innovation and **productivity** plans for operational and facilities management are achievable.

   (e) Operational **leaders** are set up for success.

   (f) Stakeholders, including but not limited to end-users, are engaged, and their feedback heard to ensure **collaboration** during operations.

   (g) Information management commitments have been upheld and **knowledge** increased through demonstrable learning and exchange of documentation.

   (h) **Risks** associated with the next stages in the life cycle are clearly identified and assessed, with suitable risk treatments in hand.

5. A suitably competent and experienced team is in place for the Operate stage. The team will include those contractors and consultants with contractual responsibilities for the early part of operations, for example for rectification of defects, and will otherwise comprise either in-house estates resources and/or subcontracted facilities or asset management resources. The Project Manager will operate in an advisory capacity in this example.

6 Validate

7 Operate: use and maintain

Purpose

The purpose of the **Operation** stage of the life cycle is to ensure the asset is used and maintained as designed, to meet the need and benefits for the client and end users and to adapt to provide best value over time. This phase is critical to validate that the design and construction process has been successful and the operational phase meets with the clients' requirements.

The eight Themes are applied in the Operate stage as follows:

- Deliver **quality** by using and maintaining in line with the design and adapting to needs over time, with clear decision criteria to justify ongoing investment in the asset. Document the detail for all adaptations and share this with all relevant operational parties.

- Uphold commitment to **sustainability** in use and maintenance in line with design criteria, including end-of-life plans. Assess how the asset meets those commitments by measuring against original criteria set out in the design. The detailed analysis will be carried out by the end user for the project manager to review.

- Identify opportunities to create additional socio-economic **value** through operation and maintenance, monitoring and managing obsolescence. This could include benefits to the environment achieved by carbon capture and also by the use of renewable energies benefitting our societies from an economical perspective. The benefits of clean energy should be measured not only from a capital expenditure perspective but from the assessment of long-term socio-economic benefit.

- Deliver **productivity** of the asset in use and maintenance in line with design, innovating to ensure continuous improvement and safe and efficient asset management by seeking the opinion of innovative technology experts.

- Establish, document and uphold clear ownership, governance and **leadership** for operations and maintenance carried out by the end user.

- **Collaborate** with stakeholders through operations, including on decisions about maintenance and end-of-life plans for the asset. Regular reviews and discussions with end users to improve performance on the current and future projects.

Code of Practice for Project Management for the Built Environment, Sixth Edition. Chartered Institute of Building.
© 2022 John Wiley & Sons Ltd. Published 2022 by John Wiley & Sons Ltd.

- Capture **knowledge** and learning on an ongoing basis and update easily accessible asset information in a controlled way, validating the integrity of changes from the design.

- Implement **risk** plans through operational activities, prioritising complex operations, resolving issues and keeping the risk process and risk register alive and current. The client project manager would be ultimately responsible for maintaining and distributing the risk management documentation if required by the end user.

Activities

The purpose and application of the themes is achieved through implementation of the following activities and creation of documented deliverables that record the issues related to each theme, via recorded minutes or by data stored in the project document management system.

The project manager will now be operating in an overview capacity. The list of activities does not represent a strict sequence of work but is indicative of a likely order. In practice, many activities will happen in parallel and with iteration in order to create the deliverables needed to approve progression to the next stage of the life cycle.

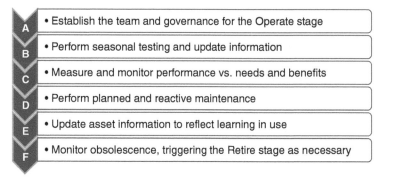

A | • Establish the team and governance for the Operate stage
B | • Perform seasonal testing and update information
C | • Measure and monitor performance vs. needs and benefits
D | • Perform planned and reactive maintenance
E | • Update asset information to reflect learning in use
F | • Monitor obsolescence, triggering the Retire stage as necessary

A. **Establish the team and governance for the Operate stage.** The **owner/operator** is responsible for commissioning a competent team to operate and maintain the asset, to ensure operational performance and to deliver the needs and benefits identified. The team may comprise of in-house or outsourced contractors. There may be continuity in the team from those involved in the design, build and handover of the asset, consistent with the soft landings philosophy (see Guidance Note 9). Contractual liabilities will exist across the defects liability period (DLP) or are associated with warranties or service-level agreements. The client will also have a liability to ensure the operation of the asset will deliver the needs and benefits identified at the start of the project. It is important that all relevant stakeholders have the necessary levels of the professional insurances related to the operational phase. Please note the need for defined roles and responsibilities and scope of works for the team, initiated by the client/operator.

B. **Perform seasonal testing and update information.** The **operator** is responsible for working with the **client project manager** to test those systems, which need to be commissioned in season, for example heating systems, air-conditioning systems, motorised natural ventilation systems. Planning for this work was conducted in the Validate stage (see Chapter 6, activity C) and should be completed as soon as practical in the Operate stage. Responsibility remains with the client until it is formally verified all seasonal systems are fit for purpose and operating effectively. The team must include people/organisations with the appropriate information management capabilities and capacities who have received a specific set of information requirements from the client. Note that information management in line with ISO 19650-3 is appointment oriented rather than project stage oriented. Any person or organisation joining the team at this stage must be issued with the specific requirements by the client. This validation is also important for insurance purposes. Please consider the timeline for the seasonal testing to ensure it is carried out at the optimum time. It is also important that proper consideration is given to the use of innovative technologies, for example in the waste-to-energy sector.

C. **Measure and monitor performance vs. needs and benefits**. The **operator** is responsible for measuring and monitoring performance as defined in the Operations and Maintenance (O&M) manual, working to ensure the asset performs as designed or to agree changed performance parameters as required. Guidance can be found in the British Standards for Facility Management[1] or Asset Management[2], where third-party accreditation of the management system is determined. The operator will have received the performance monitoring system and baseline data from the client during handover, and this is the foundation for the ongoing management of the asset. An assessment of performance in use can include:

Business objectives

- The achievement of business case objectives
- Whole-life costs and benefits against those forecast
- Continued compliance with the business strategy
- Improved operations
- The resilience of the development to business change
- Business and user satisfaction (including staff and user retention and motivation)

Design evaluation

- The effectiveness of space planning
- Aesthetic quality
- The standards of lighting, acoustic environment, ventilation, temperature and humidity
- Air-pollution and air quality
- Energy and water usage
- Maintenance and occupancy costs
- The balance between capital and running costs

---

[1] British Standards Institute (2018) BS EN ISO 41001:2008 Facility management. Management systems. Requirements with guidance for use. British Standards Institute.

[2] British Standards Institute (2014) "BS ISO 55001:2014 Asset management — Management systems — Requirements". British Standards Institute.

7 Operate

- An assessment of whether the development is being operated as designed.

- Environmental and energy consumption in use in line with relevant standards, for example ISO 14001:2015[3] and ISO 50001:2018[4]

See also GN 59 for advice on planning and implementing a process for post-occupancy evaluation of buildings.[5] Facilities and asset management services are greatly enhanced by the availability of digital technologies to allow data on the performance of the asset to be continually captured, stored and analysed. The use of big data sets greatly enhances the identification and prioritisation of operational risks, enabling proactive identification of trends to anticipate future potential operational problems, for example diagnosing potential power outages or equipment breakdown before they occur.

D. **Perform planned and reactive maintenance**. The **client** or **operator** is responsible for asset inspections, and planned and preventative maintenance, in accordance with the Operations and Maintenance (O&M) manual, to ensure the continued safe and efficient use of the asset. Planned maintenance activities will account for the condition of the asset, the criticality of the asset to the core business and the consequences of failure. Scheduling and resourcing of planned maintenance should account for any disruption the activity will cause to the organisation and occupants and should ensure resources (people and spares) are in place to complete these tasks efficiently and effectively.

Conducting maintenance activities to standard operating procedures and under risk assessments provides this reassurance that systems are working efficiently. Cyclical asset inspections are integral to the maintenance regime and should rate the asset relative to function and condition. This information then informs further investment in refurbishment works, for example window renewals or signage replacements, or to amend any maintenance agreements and/or spares inventory policies to address information about components with a propensity to malfunction. Information to be produced during each routine maintenance visit could include the date of the visit and scope of the maintenance, details of any parts replaced, details of the ongoing condition of the equipment.

In addition to planned and preventative maintenance, the client or operator is responsible for responding to day-to-day repairs and emergencies, which impair the functionality or negate compliance of the asset. It is important in the time period following handover to uphold rectification (defects liability) periods in contracts and to be clear about any guarantees or paid warranties which extend beyond rectification periods, for example a water heater with a two-year guarantee requiring replacement shortly after a one-year defects liability period. The item should be replaced free of charge by the manufacturer, but labour costs for fitting will be incurred, ideally to the original contractor for continuity purposes. Reactive maintenance, where the task is not planned in advance, may introduce risk to both safety and business continuity. Safe systems of work must be in place, including safe working practices

---

[3] International Standards Organization (2015) "ISO 14001:2015 Environmental management systems — Requirements with guidance for use". International Standards Organization.

[4] International Standards Organization (2018) "ISO 50001:2018 Energy management". International Standards Organization.

[5] Guidance Note 57: Post-occupancy evaluation of buildings

(see Guidance Note 54). Maintenance activities, whether planned or reactive, will trigger new or updated asset/facility information (see Guidance Note 10). For example, information to be produced during the unexpected replacement of a component could include manufacturer and warranty details of the new component, serial number, date of installation, maintenance requirements, expected life of the component.

E. **Update asset information to reflect learning in use.** The **client** or **operator** is responsible for ensuring asset information is up to date, documenting any modifications and maintaining a clear audit trail of decisions and changes. The core purpose of a trusted information model, maintained over time, is the 'golden thread' of information accessible to the client and occupants and prevents the need for continued access to the individuals and organisations involved in design and construction. Access to the original designers and/or builders to ensure that the integrity of the design is upheld in terms of quality, sustainability and socio-economic value should be by exception. The Operations and Maintenance (O&M) plans handed to the operator must document the client intentions for maintaining the asset, including a forward maintenance plan and associated life-cycle investment. Where this premise is challenged by experience in use, the operator is responsible for agreeing changes with the client in a controlled way across an asset on multiple occasions and this should trigger a re-evaluation of the remaining expected life of those other items and planned replacement where necessary.

F. **Monitor obsolescence, triggering the Retire stage as necessary.** The **client** or **operator** is responsible for tracking the fitness for purpose of the asset and any widening obsolescence gap[6], alerting the **client** where the asset's fitness for purpose is tracking behind the original assumptions made for its working life during design and build, or to address any changes in the client's needs and requirements. Maintenance prevents faster-than-planned obsolescence. Where planned and reactive maintenance can no longer keep an acceptable gap between business or market needs and performance (measured by benefits identified at the start of the project), the client is responsible for agreeing the relevant course of action, i.e. to operate and maintain the asset by investing further capital expenditure to assure compliance, to trigger a new project, to change use, or to sell, dispose of or demolish the asset. Refurbishment managed as part of the Operate stage, and which does not change use, may be a project following the guidance in this *Code of Practice*. Refurbishment projects serve to extend the working life of the existing asset. More significant work is needed to change the use of the asset (converting retail space into luxury apartments for example), to demolish or dispose of it, and this would trigger the Retire stage of the life cycle. In many cases where work to re-purpose/re-habilitate the asset is needed, the client may sell the asset to an organisation that is better placed to create value from such an investment. An alternative scenario is where the client decides to 'moth-ball' the asset for a period of time – ceasing beneficial use, but nevertheless requiring a duty of care to keep the asset safe and secure. Whatever the client's decision at the end of life of the original asset, they remain responsible for upholding legal requirements, and contemporaneous standards on sustainability and corporate social responsibility.

---

[6] Guidance Note 58: Monitoring obsolescence

Decisions

Before approving the project to move on from this stage, the relevant decision-makers/stakeholders are required to confirm:

1. The asset has reached a point on the obsolescence curve (see Guidance Note 60) where further investment cannot be justified and the asset needs to be retired.

2. Completeness of asset information, including the functional and condition ratings, and feedback from end users to inform future decisions in the Retire stage.

3. A suitably competent and experienced team is in place for the Retire stage, for example, to change use, demolish, dispose or sell the asset.

8

Retire: repurpose or demolish

Purpose

> The purpose of the **Retire** stage of the life cycle is to ensure there is a clear rationale to justify the end-of-life plans for the asset – either to change use/re-purpose, demolish or otherwise dispose or sell the asset to a third party and to manage this appropriately.

The eight Themes are applied in the Retire stage as follows. It is recommended that information about their application is documented and shared with known project stakeholders, in an easily accessible repository.

- Ensure plans are robust and in line with **quality** commitments in the original design.

- Uphold **sustainability** commitments in the original design, including sustainable end-of-life plans. Document the original design intent and share with stakeholders.

- Identify the rationale for next steps and the net socio-economic **impacts** this action creates.

- Identify any opportunities for innovation or to ensure **productivity** of the asset retirement plans. Agree and prioritise viable opportunities.

- Establish and uphold clear ownership and governance for the transfer of the asset from Operations into a new project with clear, concise terms of reference and roles and responsibilities.

- **Collaborate** with stakeholders to appraise them of retirement plans, gather and document their future needs.

- Capture **knowledge** and learning on an ongoing basis and update asset information in a controlled way, ensuring all information is current and complete.

- Identify, evaluate and document **risks** associated with the transition from operations to a new project.

Activities

The purpose and application of the themes is achieved through implementation of the following activities and creation of documented deliverables.

Code of Practice for Project Management for the Built Environment, Sixth Edition. Chartered Institute of Building.
© 2022 John Wiley & Sons Ltd. Published 2022 by John Wiley & Sons Ltd.

The list of activities does not represent a strict sequence of work but is indicative of a likely order. In practice, many activities will happen in parallel and with iteration in order to create the deliverables needed to approve progression to the next stage of the life cycle.

A	• Establish the team and governance for the Retire stage
B	• Agree recycling strategy
C	• Document disposal requirements
D	• Consolidate information on performance of the retired asset
E	• Handover information to the relevant client sponsor for the project to retire

A. **Establish the team and governance for the Retire stage.** The purpose of this stage is to assist the client stakeholders in the decision-making process. It will not be possible to define a team and outputs at this stage of the asset life cycle; it is about considering the options. It is acknowledged that far too many construction projects remain derelict, unused or a hindrance to their local communities, and consideration should be given in a feasibility process about whether to retire, repurpose or demolish the construction project. The **client sponsor** would be responsible for commissioning a team to plan the retirement of the current asset and to provide evidence to support the justification of the decision made. This stage is focused on confirming a strategy. Engagement of stakeholders including end users, funders and/or regulators will be necessary as part of the decision-making process. There may be multiple options, and the option selection process described in Guidance Note 16 may be useful, which outlines the considerations that need addressing when weighing up and choosing between options. The team may comprise people and organisations with skills in design or demolition, or accountants and lawyers to support a sale. Information management at this stage should be in accordance with the ISO 19650 series of standards (see Guidance Note 10). Whether the details of ISO 19650-2 or those from ISO 19650-3 are followed will depend on whether the work is being treated as a stand-alone project or as part of the operational activities. Any person or organisation appointed must be provided with specific information exchange requirements by the client. The team must include people/organisations with the appropriate information management capabilities and capacities who have received a specific set of information requirements from the client. Ideally, there is continuity of best practice in the team from those involved in the operation and maintenance of the asset, whether a third-party facilities management company or an in-house asset management team.

B. **Agree recycling strategy.** The **client sponsor** is responsible for working with the team to identify a strategy for a circular economy. The original asset may no longer be fit for purpose without substantial investment; however, there may be parts of the asset that can be re-used or re-purposed. Examples might include: re-use of some fixtures and fittings with a school or hospital setting, re-use of materials from a road surface in the mix for a new surface, re-use of brickwork following demolition of a factory chimney, upcycling or recycling of materials or fittings to create decorative features in a new build.

It is also worthy of note that the use of recycled materials could lead to those projects receiving BREEAM and LEED credits for the new client.

C. **Document disposal requirements**. The **client sponsor** is responsible for ensuring the disposal of those parts of the asset that cannot be recycled are clearly documented. These may have been specified in the original design. Where this is not the case, disposal routes must be identified and costed as input to the business case for the new project. Note that in the language of mergers and acquisitions, the sale of the asset would be termed a 'disposal'. This is not to be confused with the disposal of contaminated materials or materials that cannot be recycled, which will need to be processed via a waste management protocol.

D. **Consolidate information on performance of the retired asset**. The **operator** or **client** is responsible for providing a complete repository of evidence to the new user, of performance and learning from the retired asset, including feedback from end users, as well as technical performance data, for example a schedule of all premature equipment failures experienced during the operation of the asset. The format of this information will be as specified to the new team in their information requirements (see Guidance Note 10). This will be invaluable to the work in the Identify stage of the new project.

E. **Handover information to the relevant client sponsor for the project to retire**. The **client sponsor** for the new project may be different from the original asset. The original client has a duty of care to pass on complete and accurate information to the new sponsor. Liability for costs of any disposal are likely to remain with the original sponsor unless there has been some contractual commitment to the contrary. This should include the Health and Safety File, which may include details of any specific demolition or dismantling sequences to be followed to prevent premature collapse, the sequence for de-stressing the post-tensioned cables in a bridge deck, for example.

Decisions

Before approving the project to move on from this stage, the relevant decision-makers are required to confirm:

1. Clear justification the asset can no longer add value for the client and end users without substantial investment, including any costs for the work to retire the asset.

2. Clarity of recycling strategy and disposal.

3. Completeness of asset information ready for handover to the (potentially new) client sponsor who is accountable for the project to either re-purpose, demolish/dispose or sell the asset.

Should the decision be made to reconstruct either by new build or refurbishment processes, then the life-cycle process would recommence at Chapter 1, Identify.

Guidance Note 1   Funding mechanisms

Depending on the context, funding for projects in the built environment may be solely by the client from reserves or borrowing, or may involve a consortium of funders and funding instruments.

This guidance note provides a summary of some of the options and considerations associated that might be used.

Funding from a single, privately owned organisation

The client executive team and their board can decide how to fund the investment over time, flexing as necessary to meet the needs of the business. Investments may be funded through borrowing from investors or the market as loans in the form of overdrafts, or capital or funds from shareholders through rights issues, venture capital or grants.

In the Identify stage of the project, the client has flexibility to start work on the project without a definitive financial plan.

Funding from the public sector

In the public sector, investment may be made directly by the Treasury as part of public sector borrowing plans. Direct government funding for projects is often dependent on a cycle of 'spending reviews', which allocates funding to government departments. Each department will then assess its priorities based on the policies in place at the time and then approve projects against such policies and funding availability.

Where the project is funded by the public sector, and in early life cycle stages before the final investment decision is made, the client sponsor may only have a clear line of sight for continuation of funding to the next significant milestone as release of further funds may be contingent on meeting criteria established in governance standards.

Funding from private investors for public sector projects

Financing mechanisms known as Private Finance Initiatives (PFIs) – alternatively called Public–Private Partnerships (PPPs) – have been popular in recent decades as a way of governments incentivising the private sector to provide the

Code of Practice for Project Management for the Built Environment, Sixth Edition. Chartered Institute of Building.
© 2022 John Wiley & Sons Ltd. Published 2022 by John Wiley & Sons Ltd.

capital to invest in public projects such as the creation of roads, schools and hospitals. When government borrowing was relatively expensive, such instruments were an effective way of funding work that would have been otherwise difficult to progress. The private sector make their return on investment by leasing the completed asset to the public sector over a fixed time frame, as well as receiving interest payments.

In the United Kingdom, current plans to fund public sector projects have funding split largely 50/50 between exclusive public funding and funding by the private sector, the proportion of which can be amended dependent on public sector requirements.[1] There are also new commercial models evolving, which, for example, are incorporating novel pain–gain mechanisms, and these may become more visible in the future. Investor returns on schemes prioritising the green agenda will also influence the structure of future commercial models.

Funding from multiple investors

Some investments in assets will be made by multiple private sector investors and may involve international investors and grant funding from, for example, the United Nations or from community or heritage funds. In such cases, the goals and agenda of multiple funding stakeholders need to be aligned and auditable systems put in place to track performance. A major goal in such situations, as is the case with all projects, is to agree the delegated limits of authority at the project level, so the work can progress without unreasonable delay while maintaining the confidence of the funding bodies. See also Guidance Note 2 on risk appetite and delegated limits of authority and Guidance Note 3 on assurance and the three lines model.

Release of funds to the project team

It is usual for the finances to fund the project to be released to the team as part of a budget. Budgets may cover the whole project or part of a project such as the next stage of work, or the next financial year. Some projects included in wider portfolios of work may receive a 'seed' budget for a specific period of time before initial benefits are realised, after which the portfolio is expected to be self-funding.

Relationship with the investment appraisal and the business case

Guidance Notes 12 and 13 cover specific information about conducting investment appraisals for projects and for the justification of projects in financial and non-financial terms in a business case. The benefits, costs and risks associated with the chosen funding mechanism need to be reflected in the investment appraisal and wider business case to ensure the funding mechanism chosen represents best value.

---

[1] Keep, M. (2020) Infrastructure policies and investment. House of Commons Briefing Paper 6594. Available at https://commonslibrary.parliament.uk/research-briefings/sn06594/ (accessed December 2020).

Guidance Note 2 Risk appetite and delegated limits of authority

In the Identify stage of the project life cycle, the client sponsor establishes governance for the project. This will be extended and refined as the project progresses, and the supply chain of consultants and contractors is commissioned, but initial governance is required to direct and control early life cycle activities.

A key aspect of establishing governance is to determine the capacity of the investing organisation(s) to take risk and the sub-set of this capacity that represents the risk appetite of the client (and any investors). This is a critical aspect of the step in the international standard for risk management (ISO 31000:2018) to 'Establishing the Management Context' for risk management. Without this focusing step, any work on risk identification, analysis and treatment is misplaced as the critical question – what objectives are at risk? – has not been answered.

Using the definitions from the International Standard for Risk Management[1] and the associated International Guide to Risk Vocabulary[2]:

Risk capacity is the amount and type of risk that the organisation is **able** to take in pursuit of objectives.

Risk appetite is the amount and type of risk that the organisation is **willing** to take in pursuit of objectives.

Risk capacity is most often related to financial capacity and the strength of the balance sheet or security of funding for a project. Risk capacity also applies to the reputation of the organisation and how much risk is the organisation able to take with the trust of customers, end users, regulators, investors, shareholders, staff or other stakeholders

---

[1] International Organization for Standardization (2018) *ISO 31000:2018 Risk Management – Guidance*. International Organization for Standardization.

[2] International Organization for Standardization (2009) *ISO Guide 73 Risk Management – Vocabulary*. International Organization for Standardization.

Code of Practice for Project Management for the Built Environment, Sixth Edition. Chartered Institute of Building.
© 2022 John Wiley & Sons Ltd. Published 2022 by John Wiley & Sons Ltd.

Risk appetite needs to be expressed as a measurable threshold to represent the tolerable range of outcomes for each objective 'at risk', and using the same units used to measure performance for that objective.

An example might be:

- A client organisation is investing in a project and has a target internal rate of return (IRR) of 12% (see Guidance Note 12 for more information on the use of internal rate of return in investment appraisals).

- 12% may be perceived as the minimum tolerable return; however, a lower IRR could actually be tolerated.

- There will always be an upper limit of return that represents the best case without investing additional funds and resources into the project.

- The range between upper and lower threshold is the expression of the appetite for risk and must fit within the capacity. Note that when contracting with suppliers, this appetite for risk must be aligned with any 'cap and collar' contractual arrangements.

The same principle applies to other objectives, for example:

- Operational and staff safety performance

- Environment, Social and Governance (ESG) performance

- Time to completion

- Satisfaction of end users

The reason for expressing risk appetite as measurable thresholds is to:

1. Make objectives explicit

2. Calibrate impact scales for risk analysis and evaluation

3. Define delegated limits of authority

Delegated limits of authority clarify the responsibilities of decision-makers in the project organisation. These responsibilities will relate to decisions about commitment of expenditure and legal undertakings, and about any other decision objectives defined.

Different client organisations will have different appetites for risk depending on their objectives and what is at stake commercially and reputationally. Similarly, different client organisations will have different policies and procedures for controlling delegated authorities, and these need to be extended to the project governance to ensure decision-making on the project remains aligned with client priorities across the life cycle.

Although it is vital that investments in projects take into account whole-life costs and consider the overall return of investment (see Guidance Notes 12 and 13 on investment appraisal and business case), it is typically more effective in practice to express risk appetite through calibrated impact scales that address design and build costs and operating costs separately.

Some organisations express exposure to risks solely in financial terms, so risks to other objectives (for example, safety, sustainability, time, user satisfaction) are expressed only in financial terms – the cost of accidents, the cost of pollution, the cost of delays, the cost of dissatisfaction).

Guidance Note **2**

Other organisations express exposure to risks using a range of financial and non-financial measures – often aligned with the benefits measures that they use to determine the value contribution of the project (see Guidance Notes 7 and 8). An example may be where the project is committed to corporate social responsibility (CSR)-type benefits.

Acknowledging the choices that organisations can make in this area, two examples are provided to illustrate different approaches.

Example 1

A project, for example a national road or rail infrastructure project, to be funded by a consortium of private sector investors has a rough order of magnitude capital cost of £1.2 bn (+/−50%). The consortium has made financial provision to fund £2.5 bn costs if necessary; this is their upper threshold. They are not expecting the out-turn capital cost to be below £1.0 bn. As a result the consortium wants to understand any single risk that would add £1 m to the base cost, and they want all risks to be analysed in financial terms, so, for example, as fines for non-compliance with regulation or as the cost of delay to schedule.

The following scale is to be used to analyse the financial impact of all risks on the project.

Impact of all risks on project financials

Impact score	Size of impact with current controls	Risk owner	Delegated authority
VHI (5)	>£1,000,000	Client sponsor	Main venture governance board
HI (4)	£500,000 to £1,000,000	Client project manager	Project board
M (3)	£250,000 to £500,000	Client project manager	Project board
LO (2)	£125,000 to £250,000	Main contractor	Local management
VLO (1)	<£125,000	Main contractor	Local management

Example 2

A mid-market private business in the logistics industry has decided to invest in a new state of the art warehousing facility with advanced handling equipment and robotics. The company hasn't invested in infrastructure for many years and the Directors are concerned about stories in the media of projects that fail to meet time, cost and quality objectives. They are also keenly aware of their legislative responsibilities and the penalties for non-compliance with safety regulations in particular. They are looking to the market for partners to help them design and build the facility on their existing land.

The key criteria for the company is speed, they could use the warehousing capacity today and have no appetite for the project taking longer than 18 months and hope to complete in 12 months based on an initial estimate from their consultants. They want to understand all the downside risks which would mean the project was not delivered in 12 months, and any upside risks which could accelerate the schedule without compromising on compliance or functionality.

They are also acutely aware of their duties as Directors, so want to understand all risks, which could result in their prosecution.

The Board would like to review risk information using the following scales:

Impact of all risks on project scheduling

Impact score	Size of schedule impact	Size of regulatory impact	Risk owner	Delegated authority
VHI (5)	>24 weeks	Prosecution of Director(s) and/or >£1 m fine	CEO	Client Board
HI (4)	12–24 weeks	£500K–£1 m fine	COO	CEO
M (3)	6–12 weeks	£250K–500K fine	Client sponsor	Client project manager
LO (2)	3–6 weeks	£125K–250K fine	Main contractor	Client project manager
VLO (1)	<3 weeks	<£125K fine	Main contractor	Client project manager

For more information on risk identification, analysis and treatment see Guidance Notes 33–35 and 38.

Guidance Note **2**

Guidance Note 3　Assurance and the three lines model

Assurance is the process of providing confidence to stakeholders that the project will meet the defined needs and benefits. This process can be carried out by independent audit specialists. Effective assurance is independent from the parties performing the work and provides a service to stakeholders in support of the management of risk and effective decision-making through the project life cycle.

The three lines model (formerly known as the 'three lines of defence') has developed from corporate governance, specifically the 8th European Union Company Law Directive, and promoted by the Federation of European Risk Management Associations (FERMA)[1] and European Confederation of Institutes of Internal Auditing (ECIIA). As a result, it is incorporated into the design of governance and internal controls for many private and public organisations. For major investments, the underpinning ideas of the three lines model are adopted when designing assurance and controls for projects in many organisations. This ensures project governance fits within the overall organisational internal controls framework.

The first line includes the people in the project team performing the project activities either as part of the client organisation, or as consultants or contractors, for example designers, contractors, quantity surveyors or maintenance teams.

They:

- Monitor processes and progress of activities as defined in the project execution plan (see Chapter 3)

- Assess risks and plan treatments within their delegated limits of authority

- Report findings to the governance board at client level

- Escalate issues and decisions beyond the defined thresholds of risk appetite (see Guidance Note 2)

The second line is the governance board at client level.

They:

- Receive inputs from the first line

- Commission assurance activities in the first line, for example project gate reviews

[1]　Ruud, F. (2019) Reflections on the three lines of defense for European Commission. Available at https://ec.europa.eu/info/sites/info/files/business_economy_euro/accounting_and_taxes/presentations/presentation_flemming_ruud_2019_en.pdf (accessed 11 January 2021).

- Monitor the effectiveness of project processes and progress

- Make decisions about risks and issues escalated from the first line

- Delegate actions to manage risks and issues to the first line

- Escalate issues and decisions beyond the defined thresholds of risk appetite (see Guidance Note 2)

The third line is the executive management of the client organisation or consortium of organisations in the case of a multi-organisational investment.

They:

- Receive inputs from the second line

- Commission assurance activities in the second line, for example audit of project financial controls

- Monitor the achievement of the project business case

- Make decisions about risks and issues escalated from the second line

- Delegate actions to manage risks and issues to the second line

The specific roles and responsibilities of three lines of defence will vary depending on the nature of the client, corporate governance, and size and complexity of the project organisation. The assurance put in place by a small client building a small facility from reserves held in their bank will be very different from a multi-investor scheme for national infrastructure. However, the principle applies in all cases. Project governance should define the specific arrangements for the project in question. It is important that robust organisational structures and terms of reference are applied to each of the lines of defence.

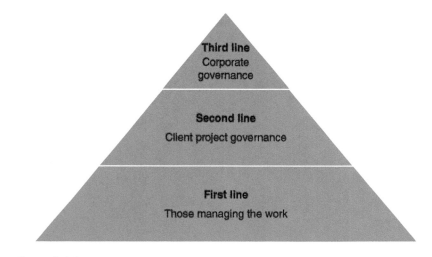

Three lines of defence.

Guidance Note 4 Design of the project organisation

There are many ways to organise project resources to deliver objectives. As the scale and scope of projects increases, the need for more specialist resources also increase. This can provide significant managerial complexity for the client organisation to manage.

The need to design the project organisation reflects the fact the client has choices to make about what work to perform in-house and what to contract, and what procurement and contracting strategy to use. More information on procurement and contracting strategy is provided in Guidance Notes 25 and 24.

This guidance note highlights some high-level options open to the client organisation as they start to shape the project.

As highlighted in Chapter 0, this *Code of Practice* emphasises the six generic and mandatory roles below:

Role	Description
End users	Occupants or users of the built environment.
Operator	Responsible for operation and maintenance of the asset as designed and built on behalf of the client and in compliance with all relevant legislation. *In some situations, the operator/ maintainer will be the same entity as the end user/occupier or may be the client in client owned and operated buildings.*
Client sponsor	Accountable on behalf of the wider client organisation, for achieving beneficial outcomes from the project including representing the needs of end users and funding bodies.
Client project manager	Responsible to the client sponsor for achieving project objectives. The project manager may be a client employee or consultant. In either case, the project manager ensures the administration of any contract(s) on behalf of the client. The client project manager should assist with support where necessary in developing a client's strategic brief, which should include CDM, establishing the project and design teams. Accountabilities should be validated and informed by the use of RACI principles through all project stages.

Code of Practice for Project Management for the Built Environment, Sixth Edition. Chartered Institute of Building.
© 2022 John Wiley & Sons Ltd. Published 2022 by John Wiley & Sons Ltd.

Role	Description
Consultant	Specialist advisors to the client team, for example architects, engineers, technology and process/method experts. Consultants may appoint their own responsible person who reports to the client project manager for the contracted scope of work. *This person may be called the consultant's project manager*. It is imperative all scopes of work for consultants are defined with specific roles and responsibilities prior to commencement of any design.
Contractor	Responsible for delivering the design, build or maintenance of the physical asset, in whole, or in part, in line with the contract(s) administered by the client project manager. Contractors may appoint their own responsible person who reports to the client project manager for the contracted scope of work. This person may be called the contractor's project manager. It is imperative all contracts are defined with specific roles and responsibilities prior to commencement of any outputs.

Professionals using this *Code of Practice*, who are representing parts of the supply chain, e.g. consultants or contractors are likely to define additional roles and responsibilities for their specific sub-set of the project, but these should always reflect the overall client context.

The charts below highlight some of the commonly used structures for the project organisation. Note that the lines represent management responsibilities and reporting lines and not necessarily contractual relationships.

Many variables exist and the three options shown are indicative, but not the only options.

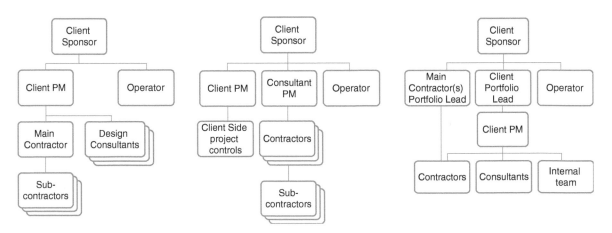

Commonly used structure for project organisation.

A more detailed example, for a complex project for instance, showing managerial, contractual and functional liaison reporting lines is shown below:

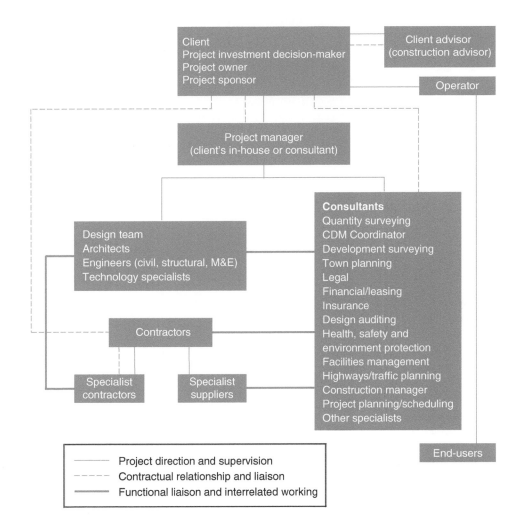

Detailed structure for project organisation.

The team clearly changes in composition as the project progresses, for example to include additional specialist consultants for technical work in design, to include dispute resolution specialists if needed during integration and handover and/or to include specialists for vertical transportation or hard landscaping. Nevertheless, there are some key aspects of forming the team that the client must ensure at every life cycle stage as outlined below:

- Definition of clear areas of responsibility and lines of authority for each project team member with deputies identified as relevant.

- Presence of clearly defined and measurable project objectives defined through performance management for the internal team, or contractually for suppliers.

- A working environment that encourages a partnership approach and an interchange of ideas by rewarding initiatives, which ultimately benefit the project.

- Ensuring that project team members are suitably located and that communication protocols have been established (particularly for electronic sharing of information) so as to facilitate regular contact with each other, as well as with their own organisations.

Notwithstanding the above tabulations and tables, it is important that due consideration is applied to the operational phase of the construction project and that similar structures with roles and responsibilities are clearly defined.

Guidance Note 5   Consenting considerations

Most projects in the built environment will require permissions to be granted by external stakeholders. Depending on the asset to be built by the project, these will include:

* planning authorities at local and potentially national levels

* regulators and related stakeholders relevant to consenting beyond planning approvals.

Planning applications and approvals will be subject to country or regional specific guidance. UK guidance can be found through the following links. Please note it is important the project manager is aware there may be continual updates on specific workflow areas such as sustainability, which they will need to monitor on an ongoing basis. The project manager needs to be looking over the horizon for potential changes that may happen related to planning.

https://www.gov.uk/planning-permission-england-wales

https://www.planningportal.co.uk/

https://www.gov.uk/guidance/community-infrastructure-levy

https://infrastructure.planninginspectorate.gov.uk/application-process/the-process/

Other permissions may be needed from, for example:

* Environment Agency, for example to protect wildlife

* Landlords content, for example for a high-street shop unit

* Centre management consent, for example a for shopping mall

* English Heritage, for example for historic buildings and places

* Highways Authority, for example any work on or near roads such as excavations, putting up scaffolding, hoarding, traditional arches or festive lighting

* Archaeological Authorities

* Local Authority consent, for example for listed buildings

Code of Practice for Project Management for the Built Environment, Sixth Edition. Chartered Institute of Building.
© 2022 John Wiley & Sons Ltd. Published 2022 by John Wiley & Sons Ltd.

- Secretary of State consent via a Development Consent Order (DCO) for Nationally Significant Infrastructure Projects (NSIP), for example for those associated with energy, transport, water and waste projects.

The examples provided are not exhaustive. The key point is that the client must identify the need for permissions and consents and engage early with the relevant stakeholders.

Guidance Note 6 Stakeholder analysis and mapping

Project stakeholders are generically defined as individual people or organisations/groups who are interested in and can affect or are affected by the project.

Understanding stakeholders and taking their views into account when planning and implementing the project is accepted as critical work in ensuring the project meets needs and benefits.

Stakeholder engagement is important for many reasons:

- To get to know and understand who your stakeholders are

- Gain an appreciation of the needs, risks and opportunities surrounding your project areas

- Explore and implement (as far as possible) the maximum benefit for the wider local communities

- To de-risk projects from the outset by being transparent across operations and aiming to deliver the best possible outcome for clients and communities

People often refer to internal and external stakeholders. Using this differentiation, internal stakeholders are people who have legal contracts with the client and are clustered around the client on the demand side (employees, customers, end users and financiers) and on the supply side (architect, engineers, contractors, trade contractors and material suppliers). The external stakeholders comprise private and public sector representatives. The private entities can potentially be from the legislative bodies, local residents, landowners, environmentalists and local pressure groups, whereas the public entities can be from regulatory agencies, local and national government and project-related third parties.

This guidance note deals with early work to analyse and map stakeholders of all types, as relevant to the specific project. Guidance Note 43 extends this to address stakeholder engagement and communication.

Various stakeholder mapping techniques can be used to identify stakeholders, their relationship to the project and with one another.

A commonly used and basic technique for mapping stakeholders is to consider the stakeholder's relative power and their interest in the project in a 2 x 2 matrix such as the one shown:

Code of Practice for Project Management for the Built Environment, Sixth Edition. Chartered Institute of Building.
© 2022 John Wiley & Sons Ltd. Published 2022 by John Wiley & Sons Ltd.

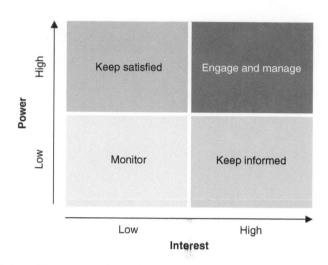

2 x 2 matrix: stakeholder power relative to interest.

Others recognise three relevant variables and map stakeholders considering:

1. Level of power and therefore potential influence on the programme

2. Level of active interest in the programme outcome, likely to be manifested as urgency

3. The degree to which the stakeholder is expected to be supportive of the project

These can be plotted either using a cube structure (all three variables) or a 'bubble-chart' type matrix with two of the variables as the axis and the third variable shown by the size of the bubble, as shown below.

Such matrices provide a visual snapshot of the current perceptions about stakeholders. These can then be validated through further engagement. To do a good job of stakeholder engagement, information about stakeholders needs to be collected and kept secure, and up to date, in line with all necessary data protection requirements.

Projects operate in a dynamic environment with the focus of the work evolving over time as the project progresses, risks emerge and learning occurs. As a

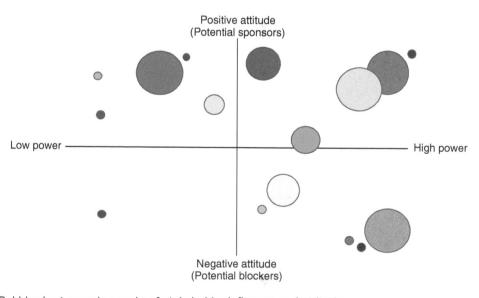

Bubble-chart mapping scale of stakeholder influence and attitude.

result, it is necessary to revisit stakeholder mapping and analysis at each project stage to revisit the landscape and ensure that false assumptions are not being made about the level of support or otherwise the project has from key stakeholders. Keeping in contact also enables projects to provide information, to canvass views and to explore potential opportunities throughout the project life cycle.

Going further than a simple mapping of stakeholders, it can be useful to understand the network of stakeholders; how they are connected and how these connections may influence their overall role, power, attitudes and potential to influence over time. It is easy to see from social networking platforms how intertwined our networks are and it is argued that everyone in the world is six or fewer social connections away from everyone else.

Social networking methods, informed by soft systems thinking, are often used to create a visualisation of the stakeholder relationships relating to a project as shown in the example in the figure. This can be valuable in shaping engagement, in recognising critical relationships and partnerships, moving beyond considerations of stakeholders as individual people or groups and preventing silos. Creating an integrated project team demonstrates not only collaboration but through the client project team, engagement ownership, leadership and partnership.

Engage with the market and senior stakeholders to consider what type of relationship is most appropriate for your project and use this to inform your choice of procurement procedure and contractual model.

Guidance Note 6

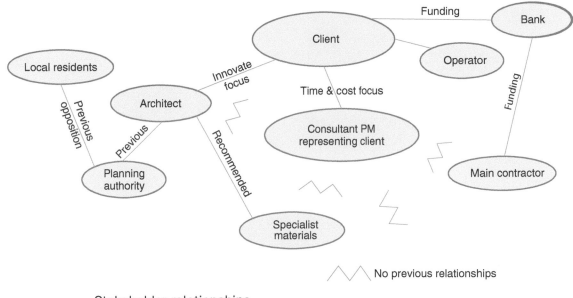

Stakeholder relationships.

Guidance Note 7   Benefit mapping

The value of an investment in a project is justified by demonstrating benefits outweigh the costs and risks. Accordingly, clients need to understand and be able to measure benefits, so they can put in place all arrangements to ensure they are realised after the project has delivered its outputs.

Established vocabulary, from a range of sources, is to refer to the outputs, outcomes and benefits of project-based work as follows:[1,2]

Output: a physical or knowledge-based asset that is produced, constructed or created as a result of a planned activity.

Outcome: the result of change, normally affecting real-world behaviour and/or circumstances.

Benefit: the measurable improvement resulting from an outcome, perceived as an advantage by one or more stakeholders, which contributes towards one or more organisational objectives.

Note: some outcomes from projects are perceived as a disadvantage by one or more stakeholders, but the project is beneficial overall and is pursued. These are often referred to as **dis-benefits** and need to be mapped and measured in the same way as benefits.

It is also important to note certain impacts that are considered as a benefit by one stakeholder may be considered as a dis-benefit by another, depending on their perception or situation in relation to the project. Such situations need to be considered carefully as part of the management of value over the life cycle.

Benefits mapping

Deciding on the benefits to measure is critical work in the early life cycle of projects. It is important benefits are not ill-defined or too generic to prevent sinking costs into projects where, should their benefits have been understood, would not have been started.

[1]   British Standards Institute (2019) "BS 6079:2019 Project management – Principles and guidance for the management of projects". British Standards Institute.

[2]   Infrastructure and Projects Authority (2017) "Guide for Effective Benefits Management in Major Projects: Key benefits management principles and activities for major projects". Available at https://assets.publishing.service.gov.uk/government/uploads/system/uploads/attachment_data/file/671452/Guide_for_Effective_Benefits_Management_in_Major_Projects.pdf   (accessed 2 January 2021).

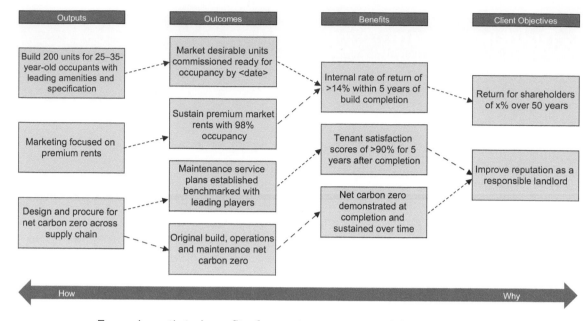

Example path to benefits for customer on a social housing project.

One way to start the thinking is to produce a benefits map (alternatively called benefits trees). Benefits mapping seeks to join up the logic and provide a coherent link between project outputs, outcomes, benefits and the value the project is intended to create.

There are two ways to start to create a benefits map:

- One is to consider the organisation objectives/needs of the project and decompose these into benefits, outcomes and outputs by asking the question 'how?'

- Alternatively, start with the project outputs and aggregate these into outcomes, benefits and organisational objectives/needs by asking the question 'why?'

In practice, the activity of benefits mapping is a messy and iterative process, which seeks to reconcile the decomposition of organisational value and aggregation of the outputs from projects. The process of creating the benefit map is often found to be as useful to clarify thinking and validate assumptions as the finished version is in communicating the project rationale.

Asking the question 'why' can create new understanding about the relative priority of needs and may, in turn, change the solution. For example, an infrastructure project may start by believing the solution is a road scheme to reduce journey times between a motorway and a port that is a popular cruise departure point. Further analysis may uncover that the priority benefits are to improve air quality in the city, a matter that if not managed would restrict port operations and therefore the opportunities for passengers and freight to use the facility. The detailed project outputs would be different depending on the benefits sought.

The example below shows the path to benefits from an investment in social housing showing not only how the investment would provide a return for shareholders but also improve their reputation as a responsible landlord.

Once completed, the benefits map makes the rationale for investments in the project clear. It provides a summary of the 'how' to realise the intended value (the 'why') and is a vital input to solution selection (Guidance Note 16) and to preparing the business case (Guidance Note 13).

Guidance Note 7

Guidance Note 8 Benefit measurement and realisation

Building from the guidance note on benefit mapping (Guidance Note 7) to realise benefits, it is necessary to have a clear measurement and a focus on realising benefits in operation.

Benefits measurement

Benefits are measurable improvements, the measure of the business outcome achieved by the project. The ability to quantify benefits and avoid claiming benefits are intangible and therefore not measurable is vital for effective project management. It is important to note, and as shown in the figure, not all benefits are necessarily measured in financial terms. How this is dealt with in the business case is covered in Guidance Note 13. The benefit map is a key communication tool to explain the nature of benefits from the project to investors and other stakeholders.

It is often the case some or all the benefits chosen use the same units of measure used to express risk appetite, for example end-user satisfaction, footfall in a retail outlet, health and safety record in operations or carbon footprint. Where different benefit measures are chosen, it is important these can be related to risk appetite, so the project does not pursue opportunities or accommodate threats that breach risk appetite thresholds.

It is best practice for each benefit to be defined explicitly, describing the benefit precisely, explaining where and when the benefit will arise, making it clear what actions are needed to realise the benefit and who will do them and defining not only how the benefit will be measured, but when the baseline measure will be established and over what timeframe the benefit will be tracked. This is commonly documented within benefit profiles, benefit registers and benefit realisation plans held by the project team and maintained throughout the project life cycle.

This discipline prevents common problems, including:

- Double-counting of benefits across multiple projects.
- Optimism bias preventing a balanced conclusion.
- Claiming benefit will be realised when the work to realise the benefit is not planned or costed.

Code of Practice for Project Management for the Built Environment, Sixth Edition. Chartered Institute of Building.
© 2022 John Wiley & Sons Ltd. Published 2022 by John Wiley & Sons Ltd.

- Setting unrealistic expectations about when benefits will arise.

- Missing opportunities to maximise the value of investments by focusing only on the asset to be delivered (the output) and not the intended socio-economic outcomes.

Defining benefits in advance of choosing the solution for the project and commencing design is vital. Benefits are the measures that will demonstrate value. Value should drive all project decisions. A focus on measurable benefits is one way that emotion and politics can be removed, or at least challenged, in decision-making.

Benefits realisation and evaluation

Defining benefits early and ensuring that design decisions take into account the impact on benefits is non-negotiable in best practice project management. Ideally, the client project manager would be accountable for benefits realisation and evaluation over the project life cycle.

Inevitably, during design there will be decisions to make to optimise the project objectives, looking at benefit/cost/risk trade-offs. Here, it is vital that the integrity of the original work on needs and benefits is maintained and that benefits, particularly those associated with social value and/or sustainability, are not traded for lower costs without revisiting the original case with the client and their key stakeholders. Legislation protects some areas, for example with the introduction of additional building safety legislation in the United Kingdom; however, there are many areas of benefit associated with projects in the built environment that can be 'value engineered' out if the original expression of benefits and associated risk appetite is not clear and reinforced. Value engineering must be focused on value and benefits, not just on creating cost efficiencies, and it should also ensure that the specification requirements are not diminished.

Benefits realisation is also a key focus during handover, commissioning, operation and maintenance of the built assets. Assets in many circumstances only provide the potential to deliver benefits in use, for example reduced maintenance costs for a school or increased well-being for housing tenants. Benefit realisation encompasses all the actions through the life cycle that prepares the operator and/or end users to use the asset in the way intended. In this respect, it is aligned perfectly with the soft landings thinking and framework.

Relevant Key Performance Indicators (KPIs) should help drive both a focus on outcomes and continuous improvement, aligning with the project performance/scorecard where appropriate, and the intended benefits to be established at procurement and realised during contract delivery.

Deciding on the correct commercial approach in line with a procurement and contracting strategy is critical to achieving the intended benefits and wider value, which should be linked to the delivery model, the desired outcomes working with the supply chain.

A focus on benefits through the life cycle ensures that the client can be confident that the project undertaken has delivered the expected value.

Guidance Note 9   Soft Landings framework

Projects in the built environment will benefit from implementing Soft Landings, which outlines duties for the client, design and construct team in order to ensure built assets are better able to meet the business, user and environmental needs. It is designed to complement rather than duplicate existing procedures used by the construction industry. This guidance note provides a summary of implementing Soft Landings.[1] The idea originated from the insights of an architect working in a building and realising the occupants were not using the asset as designed. The design concepts had been lost along the journey to create the project, and the users interacted with it in a way that the design team had not foreseen. At the same time, the industry was measuring the performance of buildings and realising that they were spectacularly failing to meet their designed energy targets.

In 2012, the UK Government created Government Soft Landings (GSL) as their client requirement for project teams to deliver better projects. A 2018 update of the framework publication reflects lessons learned by the industry utilising the Soft Landings framework. The framework provides core principles, procurement specifications and proposed activities for the different life cycle stages of the project to ensure (1) designers and the end users engage more effectively, (2) needs of end users/occupiers are not lost and (3) the asset is successfully completed. The UK government now mandates the Soft Landings approach and supports the UK BIM Framework within its Construction Playbook.[2]

Purpose of Soft Landings

Too many built assets are put into service without full commissioning or fine-tuning. Design and build teams rarely assess how their buildings perform, whether or not they meet the needs of users and management and how they can be improved. Soft Landings is a methodology to ensure operational needs are fully considered and appreciated at the design stage and embedded in procurement and contractual obligations. This is not just about better commissioning, fine-tuning and handover but to ensure greater collaboration with end users and operators from the start of a project.

---

[1]   Agha-Hossein, M. (2018) "BG54 Soft Landings Framework: six phases for better buildings". Building Services Research and Information Association.

[2]   HM Government (2020) "The Construction Playbook: Government Guidance on sourcing and contracting public works projects and programmes". Available at https://www.gov.uk/government/publications/the-construction-playbook (accessed 16 December 2020).

Soft Landings can be used for new construction, refurbishment and is designed to run alongside all forms of procurement. It introduces the concept of post-occupancy evaluation, so the project team can improve building performance and make it sustainable over the long term. Soft Landings provide for professional aftercare beginning with seasonal commissioning, graduated handover, through to operational readiness, and onto long-term, sustained high performance.

Principles embedded in Soft Landings guidance

The Soft Landings guidance includes a 'how to procure' guide, which supports the industry in procuring Soft Landing specific services from their construction supply chain. It starts with the client requirements and feeds through to contracts with the design team, contractors and sub-contractors. Soft Landings guidance emphasises the importance of delivering a design that meets the project brief while ensuring it is straightforward to use and equipped with systems that are manageable, easy to run and maintain and reduce the financial burden of operating the asset. There is a list of suggested requirements at each stage of the project to achieve this and also a set of core principles, which are:

1. Adopt the entire process
2. Provide leadership
3. Set roles and responsibilities
4. Ensure continuity
5. Commit to aftercare
6. Share risk and responsibility
7. Use feedback to inform design
8. Focus on operational outcomes
9. Involve the operator
10. Involve the end user
11. Set success criteria and performance targets
12. Communicate and inform

Six-phase approach

Soft Landings Framework 2018 is a six-phase approach designed to help the project team focus more on the client needs throughout the project, each phase has activities designed to smooth the transition into use and to address the widespread issues of assets that do not perform well during operation.

SOFT LANDINGS PHASES	Phase 1 – Inception and briefing	Phase 2 – Design	Phase 3 – Construction	Phase 4 – Pre-handover	Phase 5 – Initial aftercare	Phase 6 – Extended aftercare and POE
SOFT LANDINGS ACTIVITIES	1.1 Understand the client's needs	2.1 Review past experience to inform the design	3.1 Provide visualisations of the key space	4.1 Review the logging of quantitative (measurable) performance targets	5.1 Aftercare team to be present on site	6.1 Organise aftercare review meetings
	1.2 Define roles and Responsibilities	2.2 Hold design reviews and reality-checking workshops	3.2 Sign off key areas	4.2 Prepare a building readiness programme	5.2 Interact with the end users	6.2 Log performance data
	1.3 Review past experience to shape the brief	2.3 Exchange information	3.3 Plan for and conduct project tours	4.3 Check commissioning records	5.3 Provide support to the facilities team	6.3 Fine-tune systems
	1.4 Define sucess criteria, performance targets and evaluation methods	2.4 Review servicing and operational economy	3.4 Arrange key supply chain events	4.4 Review maintenance contract	5.4 Communicate with the end users	6.4 Record fine-tuning and usage change
	1.5 Plan for intermediate evaluations	2.5 Set requirements for tender documentation and evaluation	3.5 Accessibility review of plant and equipment	4.5 Train the future operators	5.5 Carry out walkabouts	6.5 Communicate with end users
	1.6 Create Soft Landings gateways			4.6 Plan the end users' migration	5.6 Record lessons learned	6.6 Carry out walkabouts
				4.7 Provide a place for the aftercare team		6.7 Measure and evaluate performance
				4.8 Compile a simple guide for end users		6.8 Post-occupancy evaluation (POE)
				4.9 Develop a hands-on guide for the facilities management team		6.9 Organise end of year review meetings
				4.10 Review the O&M manual		6.10 Record lessons learned

Soft Landings framework 2018 (source: *Soft Landings Framework 2018 (BG 54/2018)*, by Michelle Agha-Hossein. © August 2018. BSRIA.).

Guidance Note **9**

Guidance Note 10   Information Management using BS EN ISO 19650 series of standards

Managing information is an integral aspect of any project. The benchmark for this is now set by the ISO 19650 series of standards, which have been progressively published since 2018.

In the United Kingdom, these ISO standards have superseded the BS 1192 and PAS 1192 standards that historically were used to define BIM Level 2, which was the UK Government's required level of information management maturity. The BIM Level 2 term has now been replaced by the UK BIM Framework, and this term is now referenced in material supporting the UK Government mandate for projects to apply information management. The UK BIM Framework is supported by an extensive suite of guidance. This is available as a set of free-to-download documents and resources on www.ukbimframework.org.

ISO 19650-2, which sets out the activities to manage information on a project, requires the client to establish a set of information management documentation and resources before any project appointments are made. These include designers, other consultants and contractors. The documentation and resources that the client has to put in place includes:

- Their high-level organisational, asset and project information requirements (see UK BIM Framework Guidance D)

- The key dates and events when information is needed from the project team

- The information standard and any procedures that they expect the project team to comply with (see UK BIM Framework Guidance E)

- Any existing information or resources held by the client that will be made available to the project team to support their work

- The Common Data Environment workflow and solutions that will be used to manage and host information during its development and delivery (see UK BIM Framework Guidance C)

- The legal protocol governing information ownership, usage and liability (see UK BIM Framework Information Protocol)

Clients can appoint a specialist to help them fulfil these activities if they do not have the capability or capacity to do so themselves. However, the client always remains accountable for the information management activities (see UK BIM Framework Guidance A).

Code of Practice for Project Management for the Built Environment, Sixth Edition. Chartered Institute of Building.
© 2022 John Wiley & Sons Ltd. Published 2022 by John Wiley & Sons Ltd.

Advice from the UK government in the Construction Playbook[1] is the UK BIM Framework should be used to standardise the approach to generating and classifying data, data security and data exchange, and to support the adoption of the Information Management Framework and the creation of the National Digital Twin. Digital twinning is understanding how the power of data can help unlock value. It is a step on a journey to digital maturity and can help organisations achieve their objectives and wider social benefits. It is a realistic digital representation of assets, processes or systems in the built or natural environment.

Information management is a term that now embraces a wide range of client-side and supply chain activities. Many of these have evolved over a number of years and are reflected in project roles such as information manager, digital construction manager, BIM manager, information coordinator, document controller. These roles may still be used under the ISO 19650 standards, but all those involved need to think carefully about how existing roles work together and may need to be adapted to the new requirements of the UK BIM Framework. Key requirements of the roles supporting information management include clear and consistent identification of files and other deliverables and rigorous version control. ISO 19650 standards expect that suitable software tools are used wherever possible to support the information management process.

The figure below shows the generic information management life cycle from 19650-1, including both project delivery and asset operation phases, and positions these in the context of broader management frameworks.

It is crucial for clients to have thought about their information purposes and whether these purposes directly drive the client's information requirements issued to the various members of the project team. It is also crucial for clients

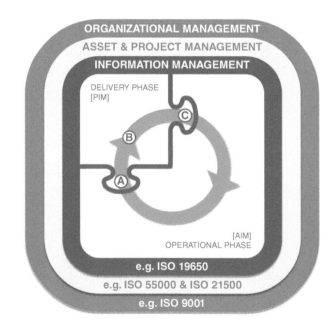

Key

A   Start of delivery phase — transfer of relevant information from AIM to PIM
B   Progressive development of the design intent model into the virtual construction model (see 3.3.10, Note 1 to entry)
C   End of delivery phase — transfer of relevant information from PIM to AIM

Project and asset information management (PIM and AIM) cycle (Source: ISO 19650-1 Figure 3).

---

[1] HM Cabinet Office. The Construction Playbook v1.0. Available at https://assets.publishing.service.gov.uk/government/uploads/system/uploads/attachment_data/file/941536/The_Construction_Playbook.pdf (accessed 17 March 2021).

to have defined how they need to receive information (e.g. what format the files need to be in, what data structure has to be used), so they can open files and import them into their own systems as easily as possible.

Specific Exchange Information Requirements (usually abbreviated to EIR) are established every time a new project/asset management team member (organisation) is appointed directly by the client. Depending on the way the project and asset teams are organised, this could be once, twice or a hundred times. Each time there is a new appointment:

- The client has to document the detailed information requirements (EIR) for that particular appointment, include these and all the background information management material in their tender/request for proposal (RFP) documentation.

- The client has to assess the responses from tenderers etc. and formally confirm the appointment (after the successful party has done some more detailed planning on how their information is going to be produced and delivered).

- Once an appointment is confirmed (agreement/contract signed), then the party has to mobilise their information management team.

- During their appointment (whether this is for design, consultancy, construction), the party produces, checks and delivers the information defined in the EIR.

- The client checks the information again before they accept it as a formal deliverable.

- Once all the information deliverables have been accepted (could be through one delivery or many depending on the nature and duration of the appointment), the information management aspects of the appointment are complete.

The initial project activities for information management and the per-appointment activities are summarised in the project information management process, which is shown in below Figure where the shaded box C represents activities for each appointment made by the client.

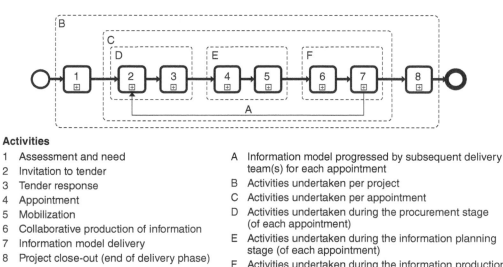

Activities

1   Assessment and need
2   Invitation to tender
3   Tender response
4   Appointment
5   Mobilization
6   Collaborative production of information
7   Information model delivery
8   Project close-out (end of delivery phase)

A   Information model progressed by subsequent delivery team(s) for each appointment
B   Activities undertaken per project
C   Activities undertaken per appointment
D   Activities undertaken during the procurement stage (of each appointment)
E   Activities undertaken during the information planning stage (of each appointment)
F   Activities undertaken during the information production stage (of each appointment)

Information management during the delivery phase of assets (Source: ISO 19650-1 Figure 3).

Guidance Note 10

The project information management process in the above figure represents activities that take place between points A and C on the above figure, so this only covers part of information management over the whole asset life cycle. ISO 19650-3 sets out a similar sequence of information management activities for the operational phase of the asset life cycle, aimed at the asset owner/operator and including any consultants, contractors or internal teams they appoint to carry out facilities or asset management activities. The interface points shown in above figure are particularly important, where the operational phase leads into project delivery (e.g. initiating a refurbishment or redevelopment) and where project delivery hands over to operation and maintenance. The transfer of information requirements and information delivery across these interfaces is set up from the operational perspective (see UK BIM Framework Guidance Part 3).

The CDM Regulations are clear that clients do have a responsibility on more complex and demanding projects to put into place management arrangements, which match the demands of the project and which extend through the life of the project. One reason why this is required is to ensure that Building Safety Managers and Facilities Managers benefit from the information generated during a construction project. With the introduction of ISO 19650, as the new standard for information management using BIM, a new tool has become apparent to help clients manage Health and Safety in their projects. This tool is the consistent adoption of Project Information Requirements (PIRs). PIRs are a class of information requirements that enable a client to set out at the start of a project, what information they regard as most important to ensure good Health and Safety performance in the project and to hand over a relevant suite of Health and Safety information at the end of the project, to support managers who have to carry out risk assessments in the building or asset that has been constructed.

The BIM 4 Health and Safety Working Group has been working with the UK BIM Framework to develop examples and guidance for clients on what a good PIR might look like, and how this might be applied in practice, integrating the PIRs with the OIRs, AIRs and EIRs in the suite of information requirements prescribed.

Transforming Infrastructure Performance: Roadmap to 2030[2] sets out a vision for innovation and reform in infrastructure delivery.

Collective efforts will be needed to address the following areas:

Improved Information Management

Effective optimisation must be founded on quality data as the key enabler for making better decisions. Data must be treated as an asset. To have quality data, we must improve organisational capability on how the information is generated, processed, managed and maintained and made available in a meaningful way for decision-makers. It is all too common that upon completion of a project, information is lost, of poor quality or not retained in a way that makes it available to inform the use of the asset, breaking the thread of information right at the very beginning.

Information Management Mandate

Set out in [annex B] is a refreshed Information Management Mandate that will be delivered through the application of the UK BIM Framework. This mandate,

---

2   Transforming Infrastructure Performance: Roadmap to 2030, https://assets.publishing.service.gov.uk/government/uploads/system/uploads/attachment_data/file/1016726/IPA_TIP_Roadmap_to_2030_v6__1_.pdf

which is applicable immediately, sets out a range of requirements for clients, including the definition of information concerning assets and projects. Public sector clients should comply with the Mandate as part of their implementation of the Construction Playbook. The Centre for Digital Built Britain (CDBB) is coordinating the ongoing work to embed awareness of the requirements of the UK BIM Framework.

Digital Twins

To enable better decisions, information must be shared across traditional sector silos. Through the creation and active management of digital twins of infrastructure, it is possible to better understand the built environment. The concept of a series of interconnected digital twins, a National Digital Twin, should unlock better overall performance through better management of the system of systems. The National Digital Twin programme is creating a system to enable this: the Information Management Framework (IMF). Initially, the programme will undertake the technical work to establish consistent meaning across data that pertains to different domains and to create the rules for effective sharing and application of data models. It is anticipated that within the next two years, as this work comes to fruition, the government will define a mechanism and timeline for incorporating the application of the IMF into a broadened Information Management Mandate.

An updated Information Management Mandate delivered through the application of UK BIM Framework.

BIM is a combination of processes, standards and technology through which it is possible to generate, visualise, exchange, assure and subsequently use and re-use information, including data, to form a trustworthy foundation for decision-making to the benefit of all those involved in any part of an asset's life cycle.

The implementation of BIM across Government projects, from delivery through to operational handover, facilitated by the adoption of Government Soft Landings 7 (GSL), has resulted in greater collaboration, productivity and efficiency in the design and delivery of construction projects delivering both social and economic infrastructure.

The report of the BIM Interoperability Expert Group (BIEG)[3] in March 2020, led in partnership by the Infrastructure and Projects Authority (IPA) and the Department for Business, Energy & Industrial Strategy (BEIS) and delivered by the Centre for Digital Built Britain (CDBB), provided evidence that interoperability is fundamental to the ability to build on the success of BIM implementation to date and to the delivery of 'whole-life' beneficial outcomes to all parties.

Interoperability, by providing a means of information transfer between different technologies while preserving the integrity of the information transferred, results in beneficial outcomes not just for clients, suppliers and end users who procure, deliver, own, operate, maintain and are served by assets. It helps to unlock societal, health and safety, environmental and economic benefits from the built environment, as well as improving its overall resilience; hence, interoperability is considered a key component of the updated BIM Mandate, now termed the Information Management Mandate.

[3] The Building Information Modelling (BIM) Interoperability Report, https://www.cdbb.cam.ac.uk/news/bim-interoperability-expert-group-report

Building Information Modelling (BIM) has been key to digital transformation, and the delivery of improved information management, across the UK built environment since the 2011 Government Construction Strategy introduced the requirement for fully collaborative BIM as a minimum by 2016. This is referred to as the UK BIM Mandate.

BIM is currently defined by the UK BIM Framework and is based on the emerging ISO 19650 series of standards and any of the remaining BS/PAS 1192 suite of standards. It was previously known as BIM Level 2 until it was superseded by the UK BIM Framework in 2018.

The Information Management Mandate is applicable immediately and for the duration of the Transforming Infrastructure Performance 10 Year Plan and requires the client to:

(a) ensure all procurement and contractual processes are compliant with the standards set out in the UK BIM Framework at the time of delivery;

(b) follow the sensitivity assessment process set out in Clause 4 of ISO 19650-5 to determine whether to implement a security-minded approach. Where a security-minded approach is required to develop and implement this following the requirements set out in ISO 19650-5 clauses 5 to 9;

(c) have in place the capability to deliver and then fulfil its information management function, as set out in ISO 19650-1, either with people within its own organisation, people acting on behalf of it or a combination of both;

(d) define its information requirements concerning its assets and projects set out in ISO 19650 Parts 2 and 3 and to support its organisational/asset whole-life and project objectives, by producing:

(i) Organisational Information Requirements (OIR)– organisational objectives;

(ii) Project Information Requirements (PIR) – purpose, design and construction of an asset;

(iii) Asset Information Requirements (AIR) – operation and maintenance of an asset;

(iv) Security Information Requirements (SIR), where applicable – in relation to information security; and

(v) Exchange information requirements (EIR) – in relation to an appointment; have a digital mechanism for defining its information requirements and then procuring, receiving, assuring and immutably storing, via a system of record, the information that it procures;

(e) fully and properly specify its information requirements, and their satisfactory delivery, within contractual documentation, recognising that the information it procures and holds is an important asset with value, that is critical to undertaking and optimising the operations, maintenance and disposal of the asset; and

(f) apply the same level of governance and rigour to the maintenance of its information, to ensure that it provides ongoing value and benefits to the client organisation. This will include the ability to share and exploit information and also make information available for regulatory purposes.

The UK BIM Framework, in addition to standards, contains BIM guidance and useful up-to-date resources that help explain the Information Management Mandate clauses listed above and which can also be used to help implement them.

The Information Management Mandate will be periodically reviewed and updated alongside the UK BIM Framework and digital and technological advancements and will include aspects of BIM interoperability. This will allow the United Kingdom to maintain its position as one of the leaders in delivering value to clients and projects through the use and management of built environment information.

Guidance Note 11 Project Mandate indicative contents

The *project mandate* is one of the critical deliverables that must be approved by the **client sponsor** and wider governance at the stage-gate review that triggers approval to move to the Assess stage. The typical content of a project mandate should include the scope of the project expressed in terms of what it will deliver, and most importantly, what it will not deliver, by clearly defining its boundaries. The layout of the mandate will be dependent on the complexity of the project. The project mandate would typically be completed by the **client project manager** for **client sponsor** approval. It forms the basis for the *project brief* in the next stage (Guidance Note 20) and in time the *project execution plan* (Guidance Note 28).

The project mandate is a key input to the business case for the project and provides the following information:

1. Document and distribution history

2. Summary of the business opportunity that triggers the project and any associated data on the imperative to act

3. Summary of the funding route(s)

4. Summary of the governance arrangements

5. Outline of needs and high-level requirements including information on the why those things are important

6. First draft of the benefits of the project showing how value will be measured

7. Outline the information management purpose and scope

8. List any existing assets in scope that will be refurbished, repurposed or retired

9. Document any assumptions or information known about timeframes.

Plus, any other information known and relevant at the Identify stage of the life cycle.

Code of Practice for Project Management for the Built Environment, Sixth Edition. Chartered Institute of Building.
© 2022 John Wiley & Sons Ltd. Published 2022 by John Wiley & Sons Ltd.

Guidance Note 12   Investment appraisal

Investment appraisal determines the economic case for the project, providing the client and any other investing organisations with information on the forecasted return on investment (ROI) of the project. It is one element important in creating the business case for the project (Guidance Note 13).

Investment appraisal uses cashflows: cash in from benefit realisation measured in financial terms and cash out – both fixed and variable capital (capex) and operational (opex) expenditure across the project life cycle. For example, for an asset to be retained by the client, roof and window renewals across a university estate will be funded by capital resources/expenditure, whereas day-to-day plumbing and electrical repairs are classified as operational expenditure. It is vital that investment appraisals take into account life cycle costs/costs in use and not just focus on the costs of the original work to create/modify the asset(s).

Projects use techniques from corporate finance, life cycle costings and cost-in-use analysis to forecast ROI using one or a combination of the following approaches:

Payback period: the length of time (usually years) for the project to achieve a cumulative net benefit, i.e. for benefits realised to equal the cash investment. Simple payback can be a useful metric for relatively short projects in stable contexts but does not take into account the time value of money.

Net present value (NPV): the amount of money, expressed in a single currency, defining what the project is 'worth' at a defined period in time when the asset is operational and realising benefits. In some cases, this will be when the asset is sold upon conclusion of the build. NPV is the sum of the 'present value' of the net cashflow in each year. The present (year 0) value of a forecasted net cashflow in year 4 is lower because inflationary influences on costs and prices would be expected between year 1 and year 4. It is critical to consider whole-life costings in calculations associated with NPV. NPV provides a way of looking at future cash flows in a consistent way, reducing bias associated with estimating likely future values. To do this, organisations agree a discount rate to apply to future cash flows.

Discount rate: the organisation's interest rate percentage representing the cost of capital for the organisation plus a further risk premium to represent the risk the investing organisation takes in using its funds on projects. Discount rates are approved as part of corporate governance and applied by projects to enable the ROI of all projects in a portfolio to be comparable.

Code of Practice for Project Management for the Built Environment, Sixth Edition. Chartered Institute of Building.
© 2022 John Wiley & Sons Ltd. Published 2022 by John Wiley & Sons Ltd.

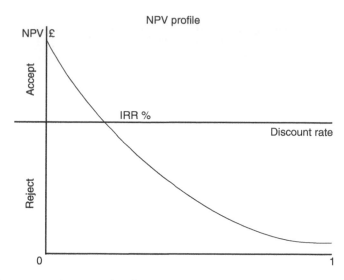

Investment case: net present value.

Weighted average cost of capital (WACC): the discount rate representing the blended cost of capital from debt and equity born by the organisations' shareholders and other stakeholders such as pension holders. The WACC represents the opportunity cost for shareholders in investing in the project and as such is used as the minimum hurdle rate for a project to achieve to be viable/have an acceptable economic return on investment.

Internal rate of return (IRR): the percentage rate of return when the NPV of cash flows is zero. IRR is commonly used as the minimum hurdle rate for projects. It differs from WACC by taking the management view of the likely returns from the project rather than a pure investor's view.

As discussed in the guidance notes on benefits, assets have non-financial value that is also considered as part of the business case. Investment appraisal refers to those costs and benefits and that can be measured in financial terms. In Guidance Note 13, which discusses the business case, consideration is given to the combined financial and non-financial justification of the project.

Guidance Note 13   Business case

The project's **business case** uses the investment appraisal described in Guidance Note 12 as a core component, but goes further in four ways to consider:

1. The value of the project to investors using the benefits identified (Guidance Notes 7 and 8) and to be measured in non-financial terms and the alignment of these benefits with the project scope.

2. The risk profile for the project beyond the risk allowances built into the investment appraisal metrics, including risks associated with effective implementation.

3. The strategic context for the project and any additional influence of this on the decision to be made.

4. The decision-making process and governance required to approve investment in a project and to ensure the project remains a good investment as it progresses through the life cycle, for example to ensure the project continues to focus on the social needs of end users, the sustainability goals of the client and/or the economic return of the investors.

Non-financial benefits

Some organisations choose to express all benefits in financial terms, so the investment appraisal becomes the single expression of return on investment. *This simplifies some things but can produce perverse results when the financial proxy for a non-financial benefit does not incentivise the desired outcomes.*

Consider this: A social housing provider is considering a project to build new homes for tenant and shared ownership occupiers over 55 years of age. The development includes communal spaces, including places to prepare and share food, with the objective of providing spaces where people can meet and support neighbours, combating loneliness and improving well-being. One desired outcome of the project is improved well-being of residents and the benefit measure chosen by the organisation is to use the annual wellbeing survey of occupiers. This measure is used across the client's estate and the project is focused on improving the well-being measures by 10% points. The client is committed to this goal irrespective of the impact on rents, or subsequent sales.

Code of Practice for Project Management for the Built Environment, Sixth Edition. Chartered Institute of Building.
© 2022 John Wiley & Sons Ltd. Published 2022 by John Wiley & Sons Ltd.

There are three ways of considering this non-financially measured benefit when considering the overall business case:

(a) See this benefit as an overlay on the investment appraisal. If the project provides an acceptable return without this benefit, then the non-financial benefit is all upside.

(b) Consider the cost of providing this benefit specifically so decision-makers can 'price' the benefit as part of their judgement of whether it is 'worth it'.

(c) Consider this benefit as the socially 'right thing to do' and commit to this aspect of the project regardless of the financial impact.

Agreeing the method to be used is a key part of agreeing governance arrangements for the project in the Identify stage of the project life cycle and is related to the investors' intentions regarding corporate social responsibility (CSR) and achieving the 'triple bottom line' making a contribution economically, morally and ecologically (profit/people/planet).

Determining the risk profile

There are a number of techniques commonly used.[1]

Risk – 'the effect of uncertainty on objectives'[2] – embraces both negative threats and positive opportunities and is a future construct. All risk assessments are estimates – educated guesses of what might happen in future made by people. As a result, they are inherently biased by a wide range of human psychological and behavioural factors that need to be recognised and taken account of where possible. One example is the effect of optimism bias on project estimates. Some organisations recognising this inherent cognitive bias have built in adjustment factors for all business cases to discount for estimator sustained false optimism.

The most comprehensive technique, often used by client organisations across the built environment, including every major government construction project and most infrastructure companies, is to build a risk model, using three-point estimates[3] for each benefit and cost-line item associated with the project representing the uncertainty caused by (1) variability, for example fluctuating productivity rates, or (2) specific risk events, for example worse than forecasted ground conditions. A simulation technique using a random number generator (typically a Monte Carlo simulation) then uses the model to predict confidence levels in different out-turns – time, financial benefit, capital, or whole-life costs, NPV, IRR. This is the most relevant approach to sizing contingency (Guidance Note 36) because it looks at the overall riskiness of the project and enables decision-makers to consider ranges not single-point estimates.

Other quantitative techniques can be used to examine parts of the investment decision, for example decision trees to compare options at a high-level, or

---

[1] International Organization for Standardization (2019) "IEC 31010:2019 Risk Management – Risk Assessment Techniques". International Organization for Standardization.

[2] International Organization for Standardization (2018) *ISO 31000:2018 Risk Management – Guidance*. International Organization for Standardization.

[3] Three-point estimates show the best/optimistic, worst/pessimistic and most likely out-turn for each activity – see also Guidance Note 35 (quantitative risk analysis).

sensitivity analysis to look at the relative sensitivity of specific risks on the overall ROI, for example how sensitive is the ROI to fluctuations in labour rates or steel prices.

Qualitative risk analysis focuses on prioritising specific risk events in terms of their likelihood and size of impact. This is useful and vital in drawing attention to specific risks as the project progresses but is less useful in supporting the investment decision or for sizing contingency. For a small project, rather than build a risk model and perform quantitative risk analysis, it is valid to assess a monetary value on each risk (likelihood x size of impact in financial terms). Adding the expected monetary values of risks is a crude and potentially misleading method, but can be useful for decision-makers in some circumstances, for example in an SME where the investment is contained in-house in terms of funding and execution.

In larger projects, building a robust view of the risk profile of a project is necessary and is specialist work. Some organisations will have in-house expertise. Others will hire specialist consultants as facilitators of the process and to provide expertise in modelling and a neutral challenge to management.

Further to the risk analysis needed to support the business case, the project must take a risk-informed view to planning all activities. This will ensure that suitable project-specific controls are approved by the Project Manager and are in place to manage the wide variety of risks that are common in projects in the built environment, including those related to health, safety, security and sustainability. This is explored in Chapter 3.

Strategic context

It is best practice for the project to have specific measurable benefits for all the outcomes the project is intended to deliver. However, in some situations there is a need to build in additional factors to give priority to specific strategic or policy areas which the project will contribute to, but where specific benefits are not directly attributable. Examples might include government policy areas such as 'achieving net zero carbon emissions by 2050'[4] or a company commitment to improving the well-being of families living in inner-city settings.

Decision-making and governance

The client sponsor is responsible for ensuring robust financial, and non-financial information is provided to inform decision-making about the affordability and achievability of the business case at each stage of the project life cycle. It would be typical for specific assurance activities to be commissioned at key decision-points to ensure the business case continues to provide an accurate representation of the strategic logic for investing in the project. For example, an audit of plans prior to a stage-gate or a peer-review of a change request prior to approval. To facilitate this process, clear ownership of risks is required (see also Guidance Note 38 – risk treatment). Further guidance in this area for

---

[4]   HM Treasury (2020) "The Green Book: appraisal and evaluation in central government". Available at https://www.gov.uk/government/publications/the-green-book-appraisal-and-evaluation-in-central-governent (accessed 17 May 2021).

publicly funded projects in the United Kingdom can be found in the *Orange Book* published by HM Treasury.[5]

Further information on estimating and the accuracy of estimates across the life cycle is provided in Guidance Note 14, and on options analysis and the decision-making process in Guidance Note 16.

Guidance Note **13**

---

[5] HM Treasury (2020) The Orange Book: Management of Risk Principles and Concepts. Available at https://assets.publishing. service.gov.uk/government/uploads/system/uploads/attachment_data/file/866117/6.6266_HMT_Orange_Book_ Update_v6_WEB.PDF (accessed 7 February 2021).

Effective project management is dependent on implementing mature methods for estimating benefits, costs and risks of the agreed business case. This is most important in early life cycle when it is also most difficult to achieve given the levels of ambiguity and variability associated with early plans. Although this is difficult, it is important to prevent commitments being made to investors and other stakeholders that cannot be upheld in practice.

Different estimation methods are relevant dependent on the degree of definition of the project and the existence of relevant historical data. It is important for the Project Manager to be aware of the difference between initial project feasibility estimates and detailed construction project estimates. For example, details regarding the initial project feasibility estimates can be found in Infrastructure and Projects Authority's 'Best Practice in Benchmarking'.[1]

Comparative (or analogous) estimates: benchmarking similar historical work to determine likely performance. This method is ideal when very similar assets have been built so benefits and costs could be expected to be comparable, such as the return from building housing stock.

Parametric estimates: using the historic relationship between key parameters and other variables to predict overall performance. This method is ideal when similar assets have been built on a different scale, but there is evidence of overall performance being driven by a few key parameters, such as the size of the main vessel in a chemical processing plant.

Analytical (or bottom-up) estimates: estimating at the level of each work package and aggregating the estimate for the whole project, for example the asset is novel and there is little or no historic data relevant to the task of estimating the project.

Over the life cycle of a project, all three methods may be used to serve different purposes.

The concept of the estimating funnel applies to projects where the accuracy of an estimate increases as more work is put into the definition of the project.

Different organisations use different terminology to describe 'classes' of estimate as projects progress through the life cycle. A generic guide is provided by

Guidance Note **14**

[1]   https://www.gov.uk/government/publications/best-practice-in-benchmarking (accessed August 2021).

Code of Practice for Project Management for the Built Environment, Sixth Edition. Chartered Institute of Building.
© 2022 John Wiley & Sons Ltd. Published 2022 by John Wiley & Sons Ltd.

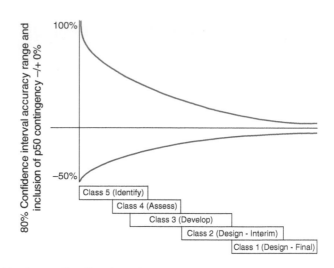

Classes of estimates: estimating funnel.

the Association for the Advancement of Cost Engineering (AACE).[2] This is supported by the UK government cost estimating guidance.[3]

The AACE approach is based on five classes of estimate as shown in the table below, applied to the life-cycle stages in this *Code of Practice*. This can be adjusted as necessary by client governance to represent the needs of the organisation. Note: these classes of estimate are most commonly applied to capital costs, but apply equally to estimates of whole-life cost, benefits and metrics used in investment appraisals, for example net present value (NPV) and internal rate of return (IRR) as described in Guidance Note 12.

Classes of estimates: five classes applied to life-cycle stages

Class	Life-cycle stage	Typical level of definition /maturity of estimate	Typical accuracy compared to target	Typical preparation effort level (dependent on type of project, amount of historical data and estimation tools and expertise available)
5	Identify	5–10%	−50 to +100%	1
4	Assess	10–20%	−30 to +50%	5
3	Develop	20–40%	−20 to +30%	10
2	Design (interim)	40–70%	−15 to +20%	20
1	Design (final investment decision)	70–100%	−10 to +15%	100

---

2   Association for the Advancement of Cost Engineering (2020) "AACE International Recommended Practice No. 17R-97 Cost Estimate Classification System". Available at https://www.pathlms.com/aace/courses/2928/documents/3802 (accessed 2 January 2021).

3   UK Cabinet Office – Infrastructure Projects Authority. Cost Estimating Guidance. Available at https://assets.publishing. service.gov.uk/government/uploads/system/uploads/attachment_data/file/970022/IPA_Cost_Estimating_Guidance.pdf (accessed 18 March 2021).

As the life cycle progresses and more work is done on estimates, the maturity of the estimate can be determined by considering:

- Inclusion of all life-cycle costs and benefits.

- The completeness of the technical baseline description defining the project and reflecting the current time-line.

- The level of detail of the scope definition (work breakdown structure) to ensure elements are neither omitted nor double-counted.

- Documentation of ground rules and assumptions.

In all cases, a well-documented estimate:

- shows the source data used, the reliability of the data and the estimating methodology used to derive each element;

- describes how the estimate was developed, so an analyst unfamiliar with the project could understand what was done and replicate it;

- provides evidence the estimate was reviewed and accepted by management.

This is specialist work and relevant expertise to support the team is required. In many organisations, this will be sourced through relationships with consulting firms with expertise in quantity surveying and commercial management for the asset type in question, although some client organisations, for example major utilities, have internal expertise and capacity in commercial departments.

Guidance Note **14**

Guidance Note 15   Materials selection

Materials selection is a strategic issue facing the built environment to address matters of performance, quality, environmental impact and whole-life costs associated with material performance, availability and the ability to re-use, recycle or dispose efficiently.

Materials selection in every project is often influenced by budget constraints, availability of materials to meet the desired timelines and the available technical expertise. It is therefore a key consideration when assessing feasibility and deciding on the chosen option for the asset.

The imperative for the built environment to use materials sustainably without compromising future use and impacting on the environment leads to the consideration of responsible sourcing.

Responsible sourcing is the practice of selecting building materials and products on the basis of a broad suite of indicators, across the sustainability spectrum and also importantly taking a holistic view of the material's provenance, i.e. from extraction, manufacturing, to its arrival on a construction site and future maintenance.

As a result, independent supply chain audits, for example to look at quarries where stone is sourced, databases of supply chain accreditation information, or responsible sourcing certification, can be used to evidence and verify a given material's provenance. In this way, the individual or firm procuring the material can be said to be sourcing responsibly – it is taking the time to investigate where materials come from, how they are made and what impact arise from this process.

The BES 6001 Framework for Responsible Sourcing standard is a commonly used scheme for the consideration of responsible sourcing in construction, this standard enables a construction product supplier to attain a third-party-verified assessment of its credentials for a given product line, and even a specific manufacturing facility. Assessment is based on the management and governance, management system standards, economic, social and environmental parameters. To achieve certification, the manufacturer has to account for a certain percentage of the constituent materials within the product.

It should be noted that responsible sourcing is distinct from Life Cycle Assessment or Environmental Product Declarations, because it also considers other, wider aspects of governance, business and social issues, such as how well the factory engages with local communities. In the United Kingdom in recent years, more attention is being paid to the Modern Slavery statements

Code of Practice for Project Management for the Built Environment, Sixth Edition. Chartered Institute of Building.
© 2022 John Wiley & Sons Ltd. Published 2022 by John Wiley & Sons Ltd.

also produced by firms, in accordance with the requirements of the 2015 Modern Slavery Act. Recent iterations of BES 6001 include a clause about Modern Slavery as an important social concern but also to recognise the risk that labour abuses can occur away from construction sites, e.g. in material factories, processing facilities and extraction sites.

The use of responsible sourcing certification for the procurement of materials can gain credits in sector schemes such as BREEAM, LEED, CEEQUAL and other environmental assessment tools.

Looking more broadly, the use of sustainable materials also offers opportunities to help the client to save money and reduce carbon emissions. For example, by considering the procurement of bulk construction materials more strategically, at an early stage in project management, it may be possible to investigate different logistical approaches, such as transport by rail or boat, which can reduce costs, pollution and carbon emissions. Importantly, getting material deliveries off local roads around a site can also reduce noise, dust and air pollution, and prevent complaints from neighbours. Just in time delivery schedules and on site manufacturing can also have a relevant impact. The regularity of deliveries can be reduced by just in time delivery and on site manufacturing and it's also possible to achieve cost and time savings by looking at using the waste reduction hierarchy – remove, reduce, reuse, recycle – aiming for resource efficiency in materials procurement. Early conversations with product manufacturers and third party logistics providers are key to maximising loads, eliminating double handling and preventing product damage and loss. Resource recovery and recycling databases offer opportunities to bring in materials unwanted on other projects in the area (or to offload them from the current site to others).

There are challenges and barriers to overcome in using sustainable materials and these by no means are limited to responsible sourcing or transportation but also includes persuading the client, educating the project team and identifying benefits.

Guidance Note **15**

Guidance Note 16 Options analysis and decision-making process

In the Assess stage of the project life cycle, a range of options need to be evaluated to check their feasibility in meeting the needs and benefits of the project, within the appetite for risk. The chosen option for the asset is approved by the client as being the most suitable in the circumstances.

Needs and benefits, and the appetite for risk were defined in the Identify stage of the life cycle and documented in the *project mandate*. These are further developed, considering specific aspects of the need and benefits desired including:

- Definition of **quality** criteria and priorities, for example compliance with specific legislation and achievement of defined levels of efficient and effective reliability measured to defined parameters.

- Definition of **sustainability** design principles, for example requirements for materials selection (see Guidance Note 15) or energy performance.

- Definition of **value** drivers and priorities, for example social impact on communities or natural habitats.

- Definition of assumptions regarding **productivity**, for example the balance of off-site/on-site construction desired.

Through this work, the objectives of the project emerge – those things that stakeholders care most about.

In considering options analysis and decision-making, to choose the way forward, there are three things to address:

1. Relative priorities of objectives – translated into prioritised decision criteria

2. The work to evaluate each option

3. The process to decide on the best option in the circumstances

Relative priorities of objectives/decision criteria

A central concept of project management is the idea that the relative priorities of objectives must be understood. Projects are constrained by time, cost and quality targets and the project manager must understand the relative priorities for the client.

Rather than just focusing on the time/cost/quality trade-offs – sometimes called the 'triple constraint' or the 'iron triangle', it is accepted that it is

Code of Practice for Project Management for the Built Environment, Sixth Edition. Chartered Institute of Building.
© 2022 John Wiley & Sons Ltd. Published 2022 by John Wiley & Sons Ltd.

Guidance Note **16**

important to understand the wider set of objectives for any single project – derived from the needs and benefits.

One technique to do this, quality function deployment (the House of Quality), provides a way to engage key stakeholders to determine which objectives are given and which can be traded, and further which objectives are complementary, or opposed.

Consider this example:

A prestige brand department store has decided to open a flagship store in a new city. Their market research has shown demand for a high-end shopping and leisure experience. There are three options for the site, in short:

1. Taking a stake in a wider investment extending the city centre shopping area to secure the premier position in the development with direct access to multi-storey parking.

2. Refurbishing a listed, centrally located building – formerly a prestigious, but now dilapidated hotel – close to the mainline railway station.

3. Refurbishing an existing department store in the high street.

Initial estimates are that all three schemes would require an investment of between £35 m and £50 m and the company is committed to compliance with all relevant standards and legislation and maintaining its exemplary record for safeguarding the health, safety and well-being of staff and contractors.

Initial conversations about needs and benefits have shown the following relationship between objectives and stakeholder priorities suggesting that of the five decision criteria, the priority order is:

1. Speed

2. Ability to service click & collect as well as traditional footfall

3. Freedom to develop and brand

4. Ability to demonstrate 'green' credentials

5. Whole-life cost and maintainability

Options analysis: relationship between objectives and stakeholder priorities

	Freedom to develop and brand	Ability to service on-line/click & collect, as well as traditional footfall	Fastest time to opening date	Ability to demonstrate 'green' credentials to stakeholders	Whole-life cost and future maintenance
Company Board	x	X	X	x	X
Legislative bodies			X	x	
Franchises that would be in store	x	X	X		
Representative customers		X	x		
In-house Facilities Management Team	x				x

Then putting the 'roof of the house' on enables relationships between objectives/decision criteria to be explored as follows:

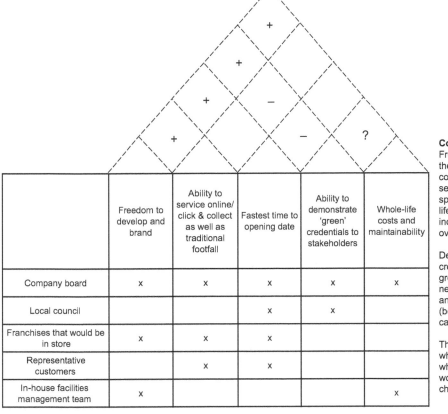

	Freedom to develop and brand	Ability to service online/ click & collect as well as traditional footfall	Fastest time to opening date	Ability to demonstrate 'green' credentials to stakeholders	Whole-life costs and maintainability
Company board	x	x	x	x	x
Local council			x	x	
Franchises that would be in store	x	x	x		
Representative customers		x	x		
In-house facilities management team	x				x

Commentary
Freedom to develop and brand the store is seen as positively correlated with the need to service click+collect, and with speed, sustainability and whole life cost/maintainability. This increases its importance overall.

Demonstrating 'green' credentials (without greenwashing) is seen as negatively correlated with speed and the click+collect desire (because it encourages use of cars not public transport).

There is a question mark about whether sustainability and whole-life cost are related – it would depend on the option chosen.

Options analysis: roof of the house to explore relationships between objectives and decision criteria

As a result, the company could decide to re-order the decision criteria as follows:

1. Freedom to develop and brand

2. Speed

3. Ability to service click & collect as well as traditional footfall

4. Ability to demonstrate 'green' credentials and whole-life cost and maintainability

Work to evaluate each option

With decision criteria made clear, each of the three options can then be evaluated.

In most cases, specialist consultants will be hired by the client to augment their internal team. In the case in point, the designated facilities management team are critical stakeholders and the client project manager needs to ensure they work well with the team of other specialists (for example lawyers, financial advisors, insurance consultants, architects, engineers, quantity surveyors, planning consultants, land surveyors) when conducting feasibility studies into each option.

Guidance Note **16**

It is important that feasibility studies of each option allow comparison of options as far as possible. Although the different options have different features, for example the condition of existing buildings will need to be assessed in two of the three cases, the purpose of feasibility studies is to inform the selection of the chosen option to meet needs and benefits.

The scope of feasibility studies will vary, but the following are areas commonly included:

- Requirements
- Health, safety and well-being across the life of the asset
- Environmental impact assessment
- Legal/statutory/planning requirements or constraints
- Timing constraints
- Cost estimates – capital and operational costs
- Risk profile
- Public support
- Other stakeholder support
- Funding options
- Geotechnical site assessments
- Brownfield conditions

The process to decide on the best option in the circumstances

Where people are involved, there will always be an element of subjectivity to the choice of solution.

Considering a minimum of three viable options detracts from a key stakeholder having a strongly held preference, which other options have to compete with.

With decision criteria ranked in advance, the choice of solution becomes more objective than if decision-makers were allowed to contribute with their own priorities foremost in their mind.

A standard decision-making process would have the following steps:

1. Identify the decision. To make a decision, you must first identify the problem you need to solve or the question you need to answer.
2. Gather relevant information.
3. Identify the options.
4. Weigh the evidence.
5. Choose among options.
6. Review and assure the decision, including commitment to action of key stakeholders.
7. Plan to take action.

Guidance Note **16**

Step 1 is covered in the Identify stage of the life cycle.

The Assess stage of the life cycle needs to cover steps 2 to 6.

The *project brief* and *intermediate business case* document the chosen solution and this is taken forward into the Develop phase of the life cycle.

Guidance Note 17   Design in early life cycle

In early life cycle and in support of options evaluation and choice of solution (Guidance Note 16), the following design considerations are relevant. The client project manager needs to manage the following:

- Set the whole team up for success with clearly defined roles and responsibilities, and a culture of positive challenge. Talk about your expectations in the team induction. Check the tender promises being made by the supplier/designer, be open about it and hold them to it!

- Develop working arrangements with the designer(s) that promote collaboration.

- Define clear project outcomes and ensure all are aligned to it – establish design principles and agree deliverables as soon as possible.

- Articulate the problem: ensure the designer is clear on the issue the project is trying to solve and the assumptions that have been made to date. Keep a watching brief on the changes base assumptions or the evidence base that assumptions are based upon.

- Fully understand the information from the previous stage. Is it clear and comprehensive? Ensure that a 'due diligence' exercise is undertaken to identify areas for further investigation.

- Understand the budget you are designing with – both for the design work and the expected whole-life costs of the asset.

- Work within your delegated time window.

- Identify and monitor your threats and opportunities and flag potential issues early.

- Make sure all existing asset data and ground information is made available to the design team; ensure the design team flags any information gaps, so they are addressed at the next stage.

- Engage with key asset stakeholders: ask for contribution, feedback and review from them.

- Involve the technical team and the principal designer in the governance process; manage the review periods proactively. If meetings help to define solutions, then organise and facilitate them.

- Ensure information is prepared in line with requirements.

Code of Practice for Project Management for the Built Environment, Sixth Edition. Chartered Institute of Building.
© 2022 John Wiley & Sons Ltd. Published 2022 by John Wiley & Sons Ltd.

Guidance Note **17**

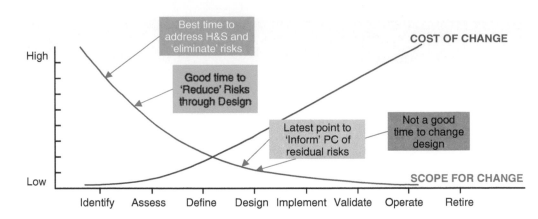

Importance of making deign decisions early.

The above emphasises the importance of making design decisions at the early design stages to influence the level of sustainability, performance and life cycle cost over the project. Therefore, as mentioned throughout this Code of Practice, design changes must always follow change control (see Guidance Note 42). Change approval processes act as gateways to ensure that all aspects of the design change request are considered before moving forward.

Guidance Note 18   Site selection and acquisition

Site selection and acquisition is an important stage in the project cycle where the client does not own the site to be developed. It should be effected as early as possible and, ideally, in parallel with the feasibility study. (It is to be noted that the credibility of the feasibility study will depend on the major site characteristics.) The work is carried out by a specialist consultant and lawyers and may involve a substantial due diligence exercise. This will be monitored by the project manager. It is, however, essential that the project manager ensures that all current and known future legislation is considered in the site selection and acquisition process.

The objectives are to ensure that the requirements for the site are defined in terms of the facility to be constructed, that the selected site meets these requirements and that it is acquired within the constraints of the outline development schedule and with minimal risk to the client.

To achieve these objectives, the following tasks will need to be carried out:

- Preparing a statement of objectives/requirements for the site and facility/ buildings and agreeing this with the client

- Preparing a specification for site selection and criteria for evaluating sites based on the objectives/requirements

- Establishing the outline funding arrangements

- Determining responsibilities within the project team (client/project manager/commercial estate agent)

- Appointing/briefing members of the team and developing a schedule for site selection and acquisition; monitoring and controlling progress against it

- Actioning site searches and collecting data on sites, including local planning requirements, for evaluation against established criteria

- Evaluating sites against criteria and producing a shortlist of three or four; agreeing weightings with the client

- Establishing initial outline designs and developing costs

- Ensure that discussions commence with the planning authorities as early on in the process as possible to validate the selection and acquisition process

Code of Practice for Project Management for the Built Environment, Sixth Edition. Chartered Institute of Building.
© 2022 John Wiley & Sons Ltd. Published 2022 by John Wiley & Sons Ltd.

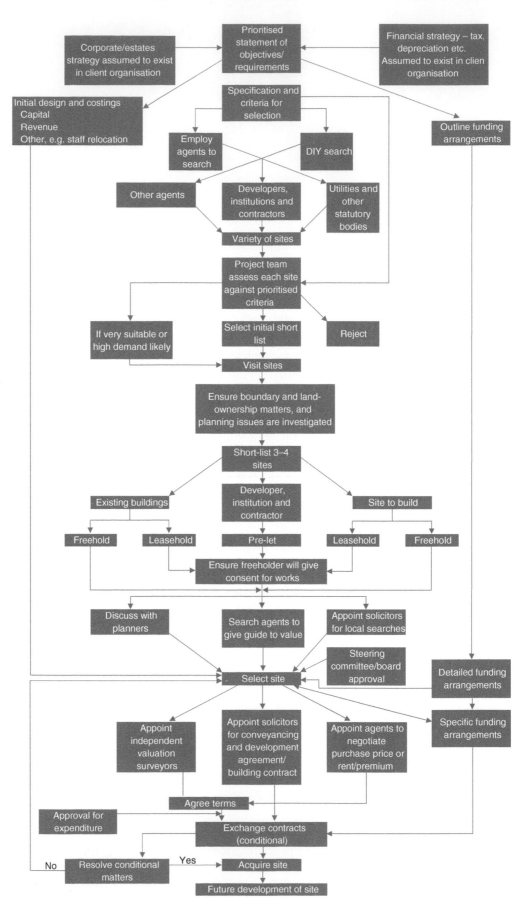

Site acquisition process.

- Ensure that all boundaries and land-ownership matters are also commenced early in the process to validate the selection and acquisition process

- Discussing short-listed sites with relevant planning authorities

- Obtaining advice on approximate open-market value of short-listed sites

- Selecting the site from a shortlist

- Appointing agents for price negotiation and separate agents for independent valuation

- Appointing solicitors as appropriate

- Determining specific financial arrangements

- Exchanging contracts for site acquisition once terms are agreed, conditional on relevant matters, for example, ground investigation, planning consent (Figure 2.4)

Guidance Note 19   Site investigations

The purpose of all site investigations is the identification of geo-technical and geo-environmental characteristics of the ground, including any contamination and live or redundant utility services, at a site or of the building or asset including its structural integrity, to provide the basis for the design and risk allocation of efficient, economic and safe projects. Early investigation allows the identification of any ground-related risks associated with a development, so they can be effectively managed and associated costs controlled. Used in this way, site investigation should be seen as an investment, which has the capacity to optimise design and hence add considerable value to a project.

The purpose of site investigations is to identify the ground conditions that may affect a development and to arrive at an understanding of the site and immediate surroundings, which will allow safe and economic development. They are a common requirement of investors and planning authorities. In the broadest sense, ground conditions are understood to include not only underlying soils and rocks but also the groundwater regime, any contamination and the effects of previous uses of the site and its environs.

The characterisation of the ground conditions whether for a 'greenfield'" or a previously developed 'brownfield' site will include consideration of both geo-technical and geo-environmental issues. Site investigations can be required for both geotechnical and geo-environmental purposes, and for many projects, it can be advantageous to combine the investigations, with resulting economies in cost, time and site disruption.

The investigation should allow a comprehensive risk assessment of the ground conditions to be made from which responses to manage or work with the risks can be developed. The risks that must be defined and dealt with may be health risks (from previous contamination), engineering risks (posed by difficult ground conditions), regulatory risks or financial risks, all of which may arise from unforeseen ground conditions and liabilities.

The client project manager must ensure that suitably qualified and experienced people are available to perform the work, for example senior geo-specialists must be professionally qualified, for example as Chartered Engineers, Chartered Geologists or SiLC (Specialist in Land Condition). Adequate levels of insurances, commensurate with the project size and risk, should be sought and provided.

The objective of the site investigation is to characterise the ground conditions. All ground-related projects have inherent uncertainties. Areas of uncertainty should be clearly identified as the project proceeds and incorporated into the

Code of Practice for Project Management for the Built Environment, Sixth Edition. Chartered Institute of Building.
© 2022 John Wiley & Sons Ltd. Published 2022 by John Wiley & Sons Ltd.

Guidance Note **19**

ground model and risk assessments. Initial uncertainties are addressed by site investigation objectives; any residual uncertainties remaining after the investigation must be identified, and possible consequences clearly stated in sufficient detail to allow safe and economic designs to be developed and to reduce, as far as possible, the occurrence and impact of unforeseen conditions.

Types of site investigation

There are many types of site investigations:

- Archaeological
- Ground
- Existing buildings or structures (i.e. asbestos, structural, condition, fire and demolition)
- Asbestos (related to demolition or refurbishment projects)
- Radon
- Contaminated land
- Unexploded bomb (UXO) surveys
- Flood risk
- Traffic and transport
- Fire hydrants
- Archaeological
- Railway and tunnels
- Utilities (electrical, gas, water and telecom-data cables)

The site investigation report (the Ground Report) must present the information obtained from the investigation in a logical and orderly format. Reports may be factual (providing the basic data) or interpretative (drawing conclusions and making recommendations) or contain both these elements. A scale plan identifying investigatory locations with relevant ground levels must be included, and this is an example of an information requirement that must have been agreed in advance of the work being performed.

A full site assessment report should include (or clearly reference) all factual information (borehole records, etc.), results of laboratory and in-situ testing, monitoring results, the conceptual ground model, risk assessments, recommendations for engineering design, appropriate remedial measures if needed and consideration of waste management issues.

Any gaps in knowledge, remaining uncertainties, residual risks and potential mitigation measures should be identified and documented by the person leading the work and communicated to the client project manager.

The approach adopted for a particular site investigation, its extent and the techniques used will all depend upon the site-specific circumstances, and the experience and judgement of the appointed team. There is no single way to carry out an investigation and inevitably different consultants will adopt different approaches for any particular project.

Guidance Note **19**

An effective site investigation depends on a clear specification of the project details by the client throughout the duration of the project. As the project progresses from feasibility study, through outline and then detailed design, and as the knowledge of the site improves, the project specification should be refined to accommodate the conditions found and the site constraints. It is common to make available to all parties, including any contractors and subcontractors for the proposed development, the results of all phases of the site investigation, including interpretation.

Phases

Phase 1 – Information gathering from available records for example maps, aerial photographs, site inspections from other companies, e.g. utilities, and from interviews and local contacts. At the end of phase 1, a hypothesis about the site is developed, e.g. about the past and present land use, geology and the surface and groundwater environment.

The walk-over survey of a site can give valuable insight into potential ground condition problems (for example slope instability or shallow groundwater) and contamination issues (revealed for example by vegetation dieback). Such site visits often give rise to anecdotal contributions by local residents. The combination of desk study and walk-over survey is an extremely cost-effective first stage in an investigation. It provides early warning of potential problems and a sound basis for the scope of intrusive investigation which is to follow. It should also identify physical and other constraints to such investigation. It provides the first information for the development of the conceptual models for the site. The conceptual ground model characterises the site in terms of its geological, hydrogeological and geotechnical conditions. The conceptual site model for potentially contaminated sites illustrates the possible relationships between contaminant sources, pathways and receptors. The desk study and walk-over survey can also provide early recognition of site issues such as ecology (for example infestation by Japanese Knotweed) and archaeology, which may have profound implications for the ability of the project to be completed in line with requirements.

Phase 2 – A limited intrusive ground investigation to gain an initial appraisal of the site and its suitability for the proposed development. This may comprise boreholes, trial pits, penetration tests, laboratory tests and geophysical methods. A factual report is usually prepared by the ground investigation contractor, and typically the appointed consultant would prepare a report setting out the conclusions and hypotheses, including the identification of problem areas that require additional investigation and consideration. The information obtained is used to design the scope of Phase 3.

Ground investigations and geo-technical reporting should be designed and undertaken by specialists in accordance with the current best practice. The design of the investigation should be based on the desk study and the preliminary conceptual ground model and also on the clearest possible understanding of the proposed development.

At an early stage, the scope of the investigation in terms of methods and strategies should be discussed and agreed with the appropriate regulatory and statutory authorities (Environment Agency, Local Authority etc.). A wide range of investigation techniques are available, and it is the responsibility of the geo-specialist to determine those best suited to particular problems and conditions that are anticipated.

It is important the ground investigation is a flexible operation, which can respond to changing circumstances as they are revealed. This is an iterative process that accumulates information in a structured way. For this reason, it is best to begin the investigation process well before other project design activities and preferably in advance of design to allow the maximum benefit to be gained.

The investigation must provide sufficient and reliable data to define the geotechnical and contamination risks, so the conceptual ground model can be revised, and qualitative and quantitative risk assessments can be made for the site.

Phase 3 – The main or detailed intrusive ground investigation, which may itself be staged. This phase of the site investigation will seek to address or clarify particular technical requirements or problem areas and to provide adequate information for design and construction. Usually a factual report on the ground investigation is prepared, and the consultant would prepare an interpretative report, providing information on the soil properties, and given design and construction guidance.

Phase 4 – Collection of information, and its appraisal, should continue during the construction works to confirm or otherwise the assumed ground model.

These four phases might run consecutively or could have long periods between them depending on the size and scale of the project, for example, is it a contained site, for example to build housing or a hospital, or multiple locations, for example to build a motorway or develop a railway.

Throughout the site investigations, the information obtained must be continuously assessed by the consultant. Changes to the proposed investigation, or even the design and construction works themselves, might be needed in the light of unexpected findings. As in all cases, changes need to be controlled using the agreed change control process (see Guidance Note 42).

It is important all parties promote a collaborative approach with clear objectives and clear responsibilities defined. Best use of information involves open communication and sharing of data at the earliest opportunity including provision of the resulting ground report to all parties involved in the subsequent works.

As is always the case, health and safety and obligations under the CDM regulations (Guidance Note 45) are important priorities in site investigation and remediation addressing matters including but not limited to: the location of unknown live services (which may be struck by borings or pits), the safe operation of site equipment, access to excavations and risks to site personnel from chemical and biological contamination, explosive gases, unstable ground and asbestos.

Production of, and adherence to, health and safety plans and auditing of health and safety procedures are essential.

The investigation itself should form the basis for safe design and construction of both temporary and permanent works.

SITE INVESTIGATION GOOD PRACTICE

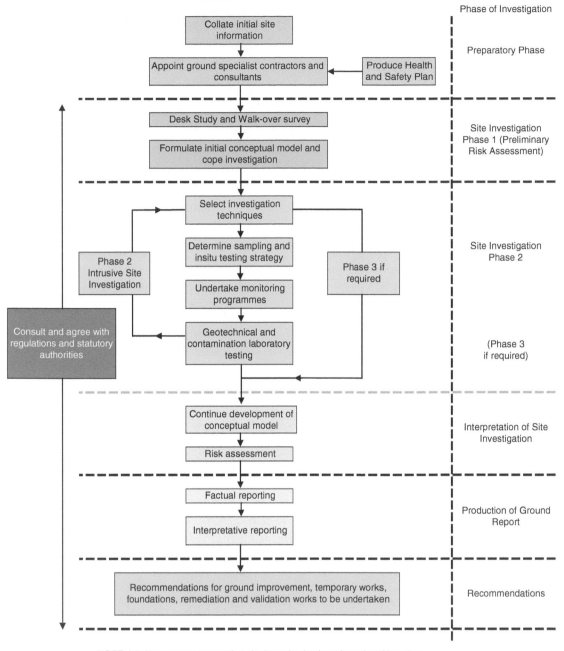

NOTE: It is important to accept that site investigation is a phased and iterative process with a requirement to review findings as they become available and to modify the investigation strategy if necessary.

Site investigation good practice.

Asbestos

Should asbestos be found, a licensed contractor should be employed and in particular, the removal of higher risk asbestos-containing materials (sprayed asbestos coatings, asbestos insulation, asbestos lagging and work involving asbestos insulating board (AIB)) should only be carried out by a licensed contractor.

The contractor will carry out an asbestos refurbishment or demolition (R/D) survey (formally Type 3 asbestos survey) on properties that are pre-2000 and are due to be refurbished or demolished (or elements of both) unless there is

Guidance Note **19**

documentary evidence that this type of survey has been conducted and all asbestos have been removed by a licensed contractor with appropriate clearance certificates etc.

Underground utility services

Underground utility surveys are an essential part of planning any construction project. Detecting and mapping out any utilities under the ground gives clarity to what can be done and where, without the need for further invasive excavations. Combining these surveys with topographical surveys will give a comprehensive map of all above and below ground.

Colour-coded utility mapping allows all involved to see the locations of pipe and cables. Things like sewers, electric cables, telecoms cables, gas and water mains will all be detected, allowing those in charge to find out who owns them to avoid any potential legal difficulties.

Knowing where the pipes and cables are can prevent any utility strikes, causing unnecessary claims, delays, disruption and potential risk of harm to workers and the general public.

Therefore, consideration should be given to the PAS 128 Specification for underground utility detection, verification and location.[1]

Unexploded ordnance (UXO) surveys

The management of Unexploded Ordnance (UXO) risk is critical to avoid delay, cost overrun and possible injury in the preparation and enabling works prior to the main construction works. The possibility of UXO being encountered on a site certainly falls within the category of a potentially significant risk. It is therefore critical that it is addressed as early as possible in the life cycle of the project. Taking the requirements of good preparation and planning, as a starting point to build upon, it is always best to appoint a specialist consultant/contractor who will follow CIRIA C681: a guide to unexploded ordnance for the UK construction industry and Unexploded Ordnance (UXO) risk management guide for land-based projects CIRIA C785.

---

[1] https://shop.bsigroup.com/products/specification-for-underground-utility-detection-verification-and-location

Guidance Note 20 Project Brief indicative contents

The *project brief* is one of the critical deliverables that must be approved by the **client sponsor** and wider governance at the stage-gate review and triggers approval to move to the Develop stage. It draws from the *project mandate* (Guidance Note 11) and forms the basis for the *project execution plan* (Guidance Note 28).

The project brief is a key input to the intermediate business case for the project and would typically address the following:

1. Document and distribution history.

2. Summary of the chosen project concept and outputs/deliverables – what work will be done in terms of designing and building new assets, or re-purposing or retiring existing assets? Make clear what work is out of scope. This will include works that will not ultimately form part of the construction contract and could include fit-out works, delivery of client specialist equipment and security requirements.

3. Summary of the information requirements for the asset and project.

4. Benefit map showing the link between outputs, outcomes, benefits and client objectives (Guidance Note 7).

5. Summary of the governance arrangements.

6. High-level project timeline (desired start and end dates, and any key milestones)

7. Resource requirements – types of resource (for example labour, materials and specialist equipment) and likely availability and costs from the market. It is important that the Project Manager monitors material costs and availability during challenging economic cycles.

8. Document assumptions and any known constraints or dependencies such as commodity availability.

Plus, any other information known and relevant at the Assess stage of the life cycle.

Guidance Note **20**

Guidance Note 21   Delivery model assessments

The term 'delivery model assessment' is the one used in central government guidance on best practice in sourcing public works projects and programmes.[1,2]

As quoted in the Construction Playbook, 'the right delivery model approach enables clients and industry to work together to deliver the best possible outcomes by determining the optimal split of roles and responsibilities'.

The core idea is a structured approach should be used, in conjunction with developing the intermediate business case (moving from SOC to OBC in government parlance) to decide which delivery model offers the best value in terms of the needs and benefits of the project.

Whole-life cost will always be a key determinant in choosing the delivery model, but other factors will also have an influence in different circumstances. For example, if the project needs to 'make the market' for future supply of suitable services with a reasonable risk profile for contractors, as is the case with the NHS hospital building programme, or if the project needs to invest in building internal capabilities rather than relying on consultants and contractors, as is the case with some private companies moving into building off-shore wind capabilities.

At the highest level, delivery model assessments result in decisions about whether to:

Potential outcomes	
	In-house – Develop own solution using internal infrastructure and expertise.
	Mixed (Make & Buy) – Components provided in-house and by external partner.
	Buy – Buy In solution from an external provider (outsource the service).
	Alternative Commercial Delivery Vehicle (JV, GovCo) – Commercial enterprise with an external partner where both parties invest in the solution. Expert advice should be sought before adopting this option.

Delivery assessment model.

1   Government Commercial Function. The Outsourcing Playbook V2.0. Available at http://data.parliament.uk/DepositedPapers/Files/DEP2020-0856/Outsourcing_Playbook_JUNE_2020.pdf

2   HM Cabinet Office. The Construction Playbook v1.0. Available at https://assets.publishing.service.gov.uk/government/uploads/system/uploads/attachment_data/file/941536/The_Construction_Playbook.pdf (accessed March 2021).]

Code of Practice for Project Management for the Built Environment, Sixth Edition. Chartered Institute of Building.
© 2022 John Wiley & Sons Ltd. Published 2022 by John Wiley & Sons Ltd.

The Construction Playbook is an example of current UK government thinking on best practice in framing and challenging thinking processes when deciding on the optimal delivery model. Summarised as follows:

Transactional approach

We know our requirement and there are people out there who can deliver so a competitive procurement process is appropriate

Hands-on leadership approach

This is going to be complex so we'll need to keep a close watch. Ensuring the outcomes meet stakeholder needs is going to be more important than lowest cost

Manufacturing approach

We are going to need lots of these over time and we'll need to progressively improve. We need to build digital and Design for Manufacture and Assembly (DfMA) capability (see also Guidance Note 23)

Hand-off design approach

We are clear on the needs and benefits but there are many ways to solve the problem. We are open to innovation and need a partner who will bring creativity and challenge as the solution may not be a physical structure

Trusted helper approach

We need help, and we know you are experienced to provide that help in our context. We need you to operate as if you are part of our team in an open, trusted relationship

The client project sponsor and manager are responsible for making sure that the right level of thinking goes into delivery model assessment and the choice of relationship with the market is aligned with the needs and benefits of the project.

Guidance Note 23   Impact of design for manufacturing and assembly (DfMA) on delivery approach

As detailed in the strategic drivers section in Chapter 0 and Guidance Note 22 on project management approaches, a delivery approach incorporating off-site manufacture and on-site assembly of some items with on-site construction for others is becoming mainstream. This is being promoted by, amongst other factors, the degree of rigidity to quality control and sustainability benefits related to off-site manufacture. Benefits to the latter point include a reduced carbon footprint by the use of less labour and reduced resource travel. Considerations related to logistical matters and interface works related to off-site construction should however be considered.

This guidance note provides some pointers on the impact of adopting a DfMA approach on the overall delivery approach of the project. The list is not exhaustive but provides a prompt list of things to consider before embarking on detailed planning.

Design for manufacturing and assembly (DfMA)

DfMA combines two methodologies, Design for Manufacturing (DfM), which focuses on efficient manufacturing, eliminating waste within a product design with Design for Assembly (DfA), which focuses on effective off- or on-site assembly, minimising resources and disruption to other activities happening on site.

Adopting a DfMA approach focuses items including:

Using common parts and materials
Reducing quantity of component parts
Reduction in carbon footprint
Simplifying part design
Designing within known capabilities for assembly
Mistake-proofing assembly (impossible to assemble incorrectly)
Reducing flexible parts and interconnections
Manufacturing for modular assembly
Adopting automated assembly where practicable

Technological advances in the renewable energy arena also offer opportunities for more modular and off-site technologies being adopted. In many of these examples, the technology needs merely to be housed and aesthetic require-

Code of Practice for Project Management for the Built Environment, Sixth Edition. Chartered Institute of Building.
© 2022 John Wiley & Sons Ltd. Published 2022 by John Wiley & Sons Ltd.

Guidance Note **23**

ments may be less apparent leading to the promotion of the use of off-site standard components.

Potential implications for delivery approach

- Design effort focused in one place – reduce potential for multiple designers to propose different parts/materials etc.

- Site restrictions need to be understood in design process, e.g. for transportation and handling and assembly of modules.

- Capabilities in on-site assembly need to be procured or developed.

- Requirement for close collaboration between designers, manufacturers and assemblers to optimise efficiency gains in both manufacturing and assembly.

- Repeat modular and off-site processes could also lead to major time benefits – a rigid off-site quality assurance protocol needs to be adopted, however, to ensure that consistent quality is being maintained – these processes should be incorporated in the relevant sections of the project execution plan.

- Tolerances and accuracy can be vastly improved in a secure and clean off-site environment. This can also include sample approval and performance testing prior to deliver to site.

- The adoption of digital technologies such as BIM can increase the efficiency and accuracy of off-site production.

- Wastage can be minimised with the maximum re-use of material in a sustainable way.

- It is important that the design responsibilities are defined and clarified between the consultant designers and the supply chain designers – and contractual documentation should reflect these clarifications.

- Fixing mechanisms between the on-site and off-site components should be carefully considered.

- Safety advantages of constructing in a more controlled environment are a benefit of off-site construction.

- There are whole-life costing benefits that can be validated by the use of off-site manufacture and components.

- Off-site construction is more predictable when the impact of weather and on-site construction delays are considered.

- More effort needs to be focused on the design, and changes to specification/method will be more difficult in an off-site environment.

Travel restriction and permits will need to be considered in the programming element related to the delivery of manufactured components as unexpected delays may occur.

Guidance Note 24   Forms of contract

In the assessment of delivery models (see Guidance Note 21), decisions will have been made about what works, goods or services need to be procured, and the initial thoughts about the relationships that need to be established in the supply chain: from transactional to partnerships.

The client needs to decide – on a continuum from transactional to collaborative – what type of relationship is desired with different suppliers. This will be guided by an understanding of what is to be procured, how critical this is to the project, the availability and reliability of the supply in the market and the risk appetite of the funders.

Adapted from Kraljic matrix[1]

For procurement strategy in general such analysis can guide decisions about, for example, frequency of tendering against the benefits of long-term relationships.

Payment mechanisms within forms of contract are a means of achieving the appropriate allocation of risk and of motivating the supplier to perform. The form of contract will drive behaviours during the project, so needs careful consideration.

In a fixed price or 'lump sum' for a defined scope, all the cost risk is with the supplier. Fixed price contracts are only appropriate where the scope of work is defined in detail, so it is appropriate to lock in expectations. The client will pay a premium to transfer the cost risk to the supplier, and only the cost risk has been transferred. Risk to other objectives (time, quality and benefits) remain with the client.

In a reimbursable or 'time and materials' contract, the client pays the supplier on an emerging cost basis and therefore carries all the risk.

Intermediate arrangements where risk is shared include 'target cost' contracts where overspend or underspend is shared between the parties using an agreed formula, or a 'bill of quantity' contract where actual quantities delivered are measured and valued against agreed unit rates. Penalties and bonus incentives are often used to motivate performance in such arrangements and such risk sharing contracts can be supported by Integrated Project Insurance. This is a form of insurance that insures project risks rather than liabilities. It operates

---

[1]   For more information on supply chain decision-making see https://www.cips.org/knowledge/procurement-topics-and-skills/supplier-relationship-management/kraljic-matrix/

Code of Practice for Project Management for the Built Environment, Sixth Edition. Chartered Institute of Building.
© 2022 John Wiley & Sons Ltd. Published 2022 by John Wiley & Sons Ltd.

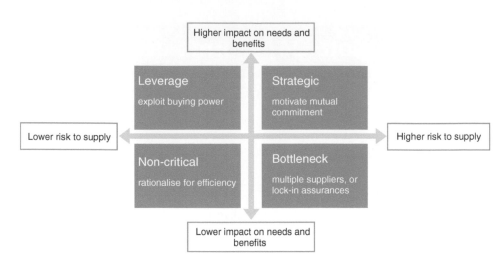

Type of relationship desired with different suppliers.

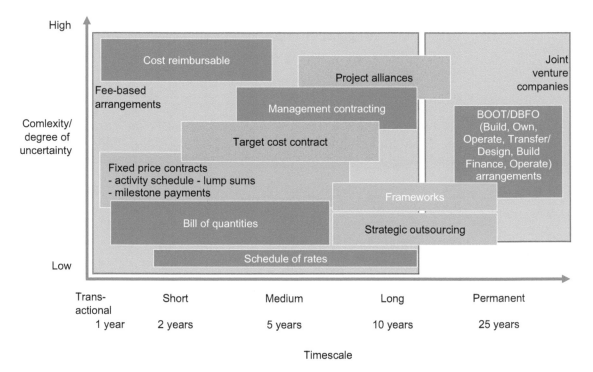

Frequency of tendering against the benefits of long-term relationships.

on a blame-free basis and insures outcomes rather than causes; all members of the project are covered – including the client – and all rights of subrogation are waived.[2] An alternative to Integrated Project Insurance is to consolidate insurance into a single policy negotiated, purchased and managed by a single sponsor, such as the client or a joint-venture operator. Typically called 'Single Project Professional Indemnity Insurance' (SPPI)[3], and aimed at large construction projects, SPPI gives the insured full control over the insurance for the project and guarantees that all the diverse interests, liabilities and risks of those involved are covered.

[2]   https://www.designingbuildings.co.uk/wiki/Integrated_project_insurance
[3]   https://www.designingbuildings.co.uk/wiki/Single_project_professional_indemnity_insurance

Some projects lend themselves to a procurement strategy where the supplier finances the development in return for a fee to operate the facility for a fixed time, for example a toll bridge.

Where longer-term projects exist, or there is a strategic need to build longer-term relationships to support a programme or portfolio, alliances such as joint-ventures can realise mutual benefits.

In all cases, the objective is to contract in a way that represents a 'win–win' situation for both client and supplier. Onerous contracts, for example where the supplier is locked into delivering but cannot make any money, are unhelpful for the sector and facilitate the 'race to the bottom' that the strategic drivers for the sector outlined in Chapter 0 seek to avoid.

Standard forms of contracts

Projects in the built environment use a number of standard forms that simplify the contracting process and provide common language and terms across the supply chain. Examples include the Joint Contracts Tribunal suite of contracts (JCT), the NEC suite of contracts and UK government Model Services Contract. Other forms of contract are available dependent on the nature and location of the works, such as the FIDIC suite of contracts.

UK Government through the Crown Commercial Service have a 'Model Services Contract' (MSC) that provides the starting point for creating a suitable contract for a wide range of public works, goods and services. Using the MSC ensures four key parameters important to government contracts. Information on this and a number of other related best practices for supply chain management for publicly funded projects can be found in the Outsourcing Playbook.[4]

Developed over decades, these standard forms of contract embed best practice in:

- Corporate Resolution Planning schedules

- Appropriate allocation of risk

- Allowable assumptions mechanisms

- Publication of Key Performance Indicators.

Framework agreements

Framework agreements (not to be confused with the NEC Framework Contract) are agreements to provide works, goods and services on predefined and speci-fied terms and conditions with a selected number of suppliers (for example consultants, designers and constructors). Entering into a framework agreement does not provide a binding agreement on either party to procure or provide, but it does specify the terms that will apply if and when they are called off. A rele-vant reference document could be Crown Commercial Services (CCS) for public works. Due to the complexity and duration of framework agreements, it is often necessary to select suitable contractors via a clear pre-qualification protocol.

---

4   Government Commercial Function. The Outsourcing Playbook V2.0. Available at https://www.gov.uk/government/news/updated-outsourcing-playbook (accessed March 2022).

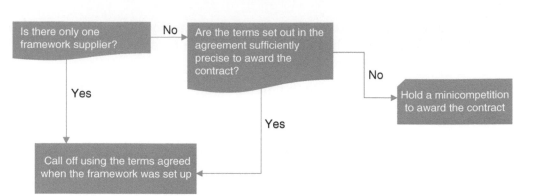

Calling off specific contracts vs a framework agreement.

This is particularly relevant in industries where there is a vast amount of ongoing construction work to procure where this work is deemed critical, for example in the water, gas, electricity and nuclear environments.

Framework agreements are common at industry level, for example the SCAPE Built Environment Consultancy (BEC) Frameworks[5] that pre-selects major consultancies for public works such as further/higher educational establishments and schools.

The flowchart below shows the process for calling off specific contracts vs a framework agreement when there is more than one supplier on the same framework.

Tendering procedures are covered in Guidance Note 26.

5   www.scape.co.uk/sectors/he-fe

Guidance Note 25   Behavioural procurement

The Theme of collaboration that runs through this Code of Practice addresses the need to build strong relationships across diverse networks of people and organisations, upholding inclusivity and equality of opportunities for work across the asset life cycle. The client project manager is responsible for creating the environment for collaboration. Collaboration alone does not secure success, but it is a critical success factor.

Procurement is often seen as the cause of poor value caused by under-performing suppliers, whereas it can and should:

- Help clients achieve better outcomes and add value to their organisation

- Ensure that everyone involved in a project wins

- Help reduce risk

- Help increase innovation

- Establish strong long-term relationships

Where it can go wrong

The detailed work of the contract is important, but so is the fit of people for the project organisation. A project team is an ever-evolving one, how a team gels together can make a huge difference, especially in how they react to adversity, and how it creates an environment that encourages innovation.

Innovative clients across the public and private sector have already seen the benefits of behavioural assessment within the tendering process as a driver for building effective partnerships, for example Highways England and the Environment Agency both attribute substantial cost savings to considering behaviours and cultural fit during procurement, making the challenge of building a high-performing team more likely during delivery. BP's approach extended to tier two and three suppliers.

Clients need to decide how early in the supplier selection process they include behavioural questions and assessments and what weighting they give to behavioural fit alongside technical considerations. The Environment Agency currently put a high weighting on behavioural and cultural fit of the client and contractor team. A recurring theme in such approaches is the value attributed to openness and transparency. This starts with the client being open and transparent about the selection process and criteria and continues throughout the relationship with a focus on collaboration, joint problem-solving and the creation and sharing of knowledge.

Code of Practice for Project Management for the Built Environment, Sixth Edition. Chartered Institute of Building.
© 2022 John Wiley & Sons Ltd. Published 2022 by John Wiley & Sons Ltd.

The use of behavioural assessment not only in procurement but throughout the onboarding and delivery process is key to ensure that the desired behaviours are embedded within the organisation (both client and supply side) and that there is continual challenge to do better.

This approach is supported by the Construction Innovation Hub's Value Toolkit[1] approach that outlines the need to agree performance measures in early life cycle to ensure that behaviours in the team and wider supply chain support the achievement of social, environmental and economic value. Defining the measures of performance in advance of optioneering, solution selection and crafting of the procurement strategy ensures that thinking about socio-economic value is built in and that incentives and reward for the whole team, including those built into contracts, are aligned.

Another example of good practice is a partnering charter signed by all parties, outwith of contractual agreements, which serves to reinforce a positive behavioural culture.

---

[1] Construction Innovation Hub (2020) "The Value Toolkit". Available at https://constructioninnovationhub.org.uk/value-toolkit/ (accessed 17 May 2021).

Guidance Note 26   Tender procedures

The overall process is summarised in the schematic below:

The tendering processes will ultimately lead to a contract award, the choice of contract should be fully analysed during the pre-construction stages of any project and any approvals to proceed with a specific form of contract should be discussed and debated at the correct governance forums.

Information handling during tender processes

Tender processes are intended to be fair to ensure that competition for contracts enables all bidders to be considered using the same criteria, eliminating bias wherever possible. Accordingly, information handling is a key part of the tendering process.

A pre-qualification process may be implemented to ensure the suppliers taking the time to tender for the work meet the basic requirements.

Areas to consider when designing a tender process include:

- Ensure all appropriate health and safety references are made in the documentation together with any relevant statutory authority references.

- Coordinate all contract documentation to ensure all requirements are covered, for example where aspects of design are involved or to confirm warranties are secured where necessary.

- Ensure method statements and reports are specified as a requirement.

- Interviewing successful supply chain applicants to clarify any special conditions and to meet significant leading personnel.

- *Arrange for formal acceptance of tendering applicants as appropriate and issuing award documentation*.

- Base selection on a balance between quality, schedule and price.

- Initiate action if price submissions are outside of the budget. It is advised at commencement of the tendering process to agree a strategy regarding the procedures for managing the possible event of the tendered price exceeding the anticipated budget.

- Ensure the client understands the nature and terms of the construction, particularly those in relation to possession and payment terms, and confirm

Code of Practice for Project Management for the Built Environment, Sixth Edition. Chartered Institute of Building.
© 2022 John Wiley & Sons Ltd. Published 2022 by John Wiley & Sons Ltd.

Guidance Note **26**

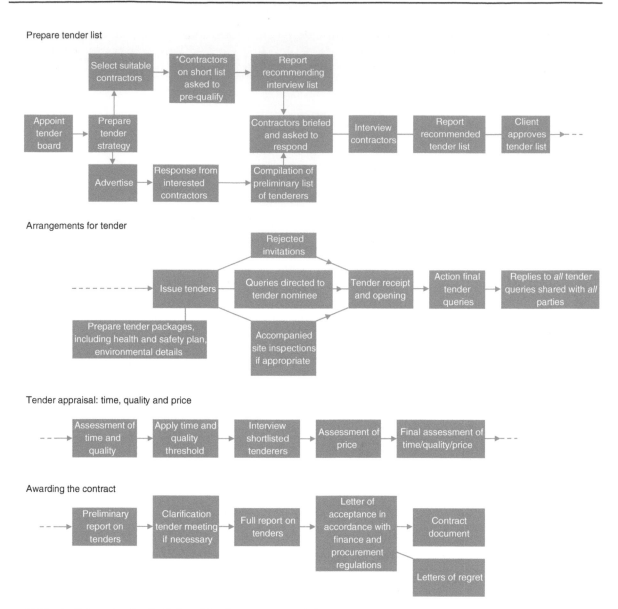

Prepare tender list

Arrangements for tender

Tender appraisal: time, quality and price

Awarding the contract

*Optional–will depend on the client's financial and procurement regulations

Tender procedure.

possession of the site can be given to the contractor on the date set out in the tender.

- Arrange for formal signing and exchange of contracts.

The client project manager is responsible for ensuring that information is produced at the appropriate time and ensuring they contain all specialist items required by the client and other stakeholders. For example, for pre-construction works such as demolition, site clearance, access and hoarding ensuring specialist terms required for activities such as archaeology and environmental investigations are included. It is likely that the client project manager will engage procurement specialists to support this work, either members of the client organisation or specialist contractors.

It is increasingly common for clients to use procurement processes enabled by electronic tools and systems to increase efficiency. This is typically referred to as e-procurement. Electronic procurement is becoming more commonplace in the construction life cycle. In effect, it converts traditional procurement into a computerised process. It is therefore critical that security measures are taken to ensure correct processes and procedures to ensure fairness are adopted. There is a view that e-procurement de-personalises the selection process and this risk should be carefully monitored.

Procurement under EU directives

EU Directives (Directive 2004/17/EC and 2004/18/EC dated 31 March 2004) enacted by UK (The Utilities Contracts Regulations 2006 and The Public Contracts Regulations 2006 enforced 31 January 2006) require that for public procurement above a certain monetary threshold, contracts must be advertised in the *Official Journal of the European Union* (OJEU) and there are other detailed rules that must be followed. This is irrespective of the procurement route chosen.

The European Union (Withdrawal) Act 2018 (EUWA) provides a new constitutional framework for the continuity of 'retained EU law' in the United Kingdom, replacing the EU treaties that had until that point applied in the United Kingdom. Thousands of amendments to that retained EU law also entered into force at the same time.

Example of relevant legislation include, but are not limited to, the UK Government guidance on public sector procurement.[1]

However, as the UK progresses away from the EU, stakeholders are requested to take advice regarding the current status of their projects related to any UK Government directives.

Generally, contracts covered by the regulations were the subject of a call for competition by publishing a contract notice in the OJEU.

For public projects, further valuable information can be found in the Construction Playbook-Government Guidance on Sourcing and Contracting Public Works, Projects and Programmes. Version 1.0 dated December 2020.[2]

The Construction Playbook is the result of extensive collaboration from across the public and private sectors to bring together expertise and best practices. It builds on the recently published National Infrastructure Strategy and supports the UK Governments ambition to transform our infrastructure networks over the next decade and beyond so that we can build back better, faster and greener.

[1]   https://www.gov.uk/guidance/public-sector-procurement
[2]   https://www.gov.uk/government/publications/the-construction-playbook

Guidance Note 27 Dispute resolution

The first step in any dispute between parties in the supply chain is to try to negotiate an agreement.

Where this is not possible, it is a well-established principle that litigation should be seen as a last resort to resolve any disputes in the supply chain. Non-adversarial alternative dispute resolution (ADR) such as mediation is encouraged and expected by any judges who preside over any cases that eventually do come to court.

Disputes may arise from a wide range of situations, including but not limited to design faults, defective work or materials, withholding of payments or loss and expense for delay.

Mediation, conciliation, expert determination, adjudication and arbitration are all possible ADR methods preferable to litigation. Depending on the approach taken, the determination may be binding or non-binding, as described below.

Non-binding alternative dispute resolution processes

Mediation and conciliation are non-binding processes where a third-party dispute resolver helps the parties to agree their differences. These are entirely private processes, conducted without prejudice to the rights of either party. If they do not succeed in reaching a settlement, there is nothing to prevent either party from dealing with the same dispute through another forum at a later date. The dispute resolver will agree a process and work with both parties separately and together to come to an agreement. Although mediation and conciliation are essentially non-binding processes, both parties can agree that final settlement should be binding, agreeing to share the costs of the dispute resolver.

Mediation

Without express permission, the mediator will never disclose what has been said by either party to the other. A mediator does not have to have a detailed understanding of the facts or the law of the matters in dispute but it often helps. The mediator does not advise the parties of their rights nor generally advise the parties of the strength of their case. Instead the mediator helps each party to see the strengths and weaknesses of the situation. In doing so, the purpose of mediation is to improve understanding with a view to executing an agreement to settle differences.

Code of Practice for Project Management for the Built Environment, Sixth Edition. Chartered Institute of Building.
© 2022 John Wiley & Sons Ltd. Published 2022 by John Wiley & Sons Ltd.

Conciliation

Conciliation as a similar process to mediation but the conciliator takes a more active role in the settlement of the dispute than does the mediator. A conciliator necessarily has to have to a detailed understanding of the facts and law of the matters in dispute. The conciliator will express an opinion on the relative merits of the parties' respective cases and attempts to guide the parties into an agreement compatible with the parties' rights under the contract.

Final and non-binding

Unlike ADR in which the parties make their own decision with the support of a mediator or conciliator, expert determination and adjudication are processes where a third party is introduced to make the decision. Because the process is consensual, it is always a private process. However, depending upon the rules of engagement agreed between the parties, the available information may not be privileged and the decision made may not be binding on the parties, leaving them free to revisit the dispute in another forum. The parties are free to agree who should pay the dispute resolver's costs and how the party's costs should be dealt with, although it is usual for each side to pay their own costs.

Expert determination

Expert determination is quite different from any other method of dispute resolution. In this forum, the expert is appointed for their knowledge and understanding of the particular issues in dispute in the field in which they are an acknowledged expert. There is usually no provision for the parties to change their position or amend their case during the process. The expert is given the role of investigator and is under no obligation to consult with the parties privately, or together, although this does often happen. The expert is required to find the facts and law in relation to the issues in dispute to make their own inquiries, tests and calculations and to form their own opinion and decide upon the merits of the parties' position.

Adjudication

In England and Wales and in several Commonwealth countries, adjudication has statutory authority. Under the law of those countries adopting this process, it is generally the rule that either party to a specified type of construction contract, as defined in statute, has the right at any time to submit any dispute or difference to the adjudication of a third party. However, even where the statutory right is limited to particular types of contract, there is nothing stopping the parties from agreeing by contract to follow the same process in regard to contracts that are outside the Act.

Adjudicators are often appointed for their knowledge and experience of the type of matters in dispute, although it is not essential. Although the idea of adjudication is there should be a decision, in the event that the parties do not like the result there is nothing to prevent them from running the case again in another forum. The adjudicator will read the parties' respective referrals and any documents provided in support. They may also require a hearing and will often conduct conference calls with the parties. The parties must have an equal opportunity to make their own case and to respond to the case against them, although they may not alter or amend their submissions

The adjudicator's decision is binding until either party decides to refer the same dispute to arbitration or litigation, in which case the decision is binding until an award or judgement is handed down.

Final and binding

The only final and binding processes are arbitration and litigation.

Arbitration

An arbitration agreement is written into all standard forms of building and civil engineering contracts. It is a private process and nobody is permitted to know of the matters in dispute or the decision unless the parties agree otherwise. The arbitrator's decision is final and binding and can be enforced in many countries of the world by virtue of the New York Convention. Arbitrators, like judges, must be independent and impartial. They must follow the law of the contract and the rules of natural justice to provide a speedy and efficient decision on all the issues submitted to jurisdiction. The arbitrator may not go outside the limitation to decide things not part of the reference.

Subject to the arbitration agreement, the parties may adopt specific procedural rules that dictate the powers of the arbitrator or the procedure to be followed, or the powers of the arbitrator are set out by statutory instrument. In domestic disputes, it is normal for the reference to be to a single arbitrator, but in international disputes it is more common for each party to appoint their own arbitrator and for the arbitrators to appoint a chairperson or umpire, forming a three-person tribunal.

Generally, the arbitrator has the powers of a High Court judge in regard to the taking of evidence on oath, subpoenas for evidence, discovery etc. They can order a party to pay the costs of interlocutory matters and can determine who should pay the arbitrator's fees and whether the losing party should pay the winning party's costs, in whole or in part, with or without interest and on what basis. The arbitrator must give reasons for their decision if either party requests it.

Litigation

Litigation is the dispute resolution process run by the civil courts of the state. It is free to every individual who has a grievance to resolve.

Litigation is a public process (justice must be seen to be done), and the public are encouraged to sit in on the proceedings to hear of the matters in dispute. Judges must give reasons for their decisions, and important decisions are published and recorded in law reports.

Guidance Note 28 Project execution plan indicative content

The Project Execution Plan (PEP) is a key information source for any project. It is the collection of up to date plans and protocols for carrying out a project that is owned by the client project manager. In some environments, this is known as the project management plan (PMP), the project initiation document (PID) or the project handbook.

The depth of information contained within a PEP reflects the complexity of the project. A small project resourced in-house by the client or using a small number of consultants working with the client project manager will contain all detailed plans to deliver the project. A large project that has the involvement of numerous large contractors, as well as consultants, may be at a higher level with the detail contained in plans maintained contractually by the supply chain. Nevertheless, the PEP and its associated contractual plans must cover as a minimum the why, what, who, how, where and when for the project.

In some cases, the client project manager may manage the creation and maintenance of the PEP within their own client team. In most cases, this activity will be contractually delegated to the party responsible for the Implement stage of the project, who produces and maintains the PEP on behalf of the whole project team, including all consultants and contractors.

The PEP is unlikely to be a single physical document (other than for the smallest projects). It is most likely to comprise a collection of documents and other information that are coordinated and together describe the complete plans for the project.

The PEP must be a live document that is baselined at the end of the Define stage of the life cycle and kept up to date through the life of the project in line with the agreed change control process (see Guidance Note 42). It is essential that the client project manager has a continual and full understanding and awareness of the plan. The plan should reflect the particular location and environment in which the construction is taking place and should be project specific and not generic.

The elements of the PEP subject to change control include:

- Statements of needs and benefits, including key performance indicators

- Scope definition/breakdown

- Quality plans

Code of Practice for Project Management for the Built Environment, Sixth Edition. Chartered Institute of Building.
© 2022 John Wiley & Sons Ltd. Published 2022 by John Wiley & Sons Ltd.

Guidance Note **28**

- Health, safety and well-being protocols for all working on the project, either on or off site.

- Other protocols, as necessary, to guide the project team on ways of working, for example sustainability plans

- Dependency networks

- Resourced time plans (schedules/programmes)

- Procurement protocols and plans

- Team structure, roles and responsibilities, delegated limits of authority/ parameters of empowerment (for example to spend, approved changes etc.) including for contractors and consultants

- Budget/cost plans

- Risk management process

- Risk analyses (qualitative risk registers and any quantitative analysis)

- Contingency plans

- Issue resolution/problem-solving records (this may include evidence from audits or other assurance activities)

- Stakeholder analysis and stakeholder engagement and communication protocols and plans

- Change control process

- Information management process, including for record keeping, meetings and minutes

- Decisions taken at stage gates and other key decision points identified in the plan.

Some project managers also choose to include progress information in the PEP. For example, each month the outputs of progress monitoring are incorporated into the plan, so it is possible to see actuals vs plans and any additional remedial actions are clearly justified.

The PEP is a vital source of project information. Elements of it are used throughout the project, for example to:

- Onboard new team members

- Communicate progress to stakeholders

- Document an audit trail of information, from risk registers to documenting decisions at stage gates to evidence of change control.

- Remind team members of roles and responsibilities, key dates and protocols.

Assure processes and procedures are being carried out satisfactorily.

Guidance Note 29   Scope and quality planning and management

Planning starts in a project from the earliest stages of the life cycle when initial top-down or benchmarked estimates of time to completion and whole-life costs are made for the high-level business case and *project mandate*. These are refined for the intermediate business case and *project brief* following selection of the chosen solution to meet the needs and benefits.

Drawing from the *project brief*, and considering the work on delivery approach and procurement strategy, the work can commence to plan the project in detail to create the plans that form part of the overall project execution plan.

The client project manager remains responsible for the adequacy and completeness of all plans for the project despite some of the detailed planning being completed by consultants and/or contractors in most scenarios.

The first step is to define the scope (what work will be done) and quality (the standards to which the work will be completed).[1] Quality encompasses regulatory compliance and related targets and values related to health, safety and well-being of the project team and of the operators and end users of the assets.

Breaking down or decomposing the scope

A project scope is the sum of the work content of a project. The definition of project scope must build from the statement of needs and reflect the benefits measuring the value created by the project.

A range of different breakdown structures are used to literally 'break-down' or decompose the project into more detailed parts. They provide an essential structure for project planning, monitoring and control. There are three primary breakdown structures commonly used during core project planning. These are product breakdown structure (PBS), work breakdown structure (WBS) and organisation breakdown structure (OBS). For some smaller/simpler projects, the use of one of these breakdown structures would be sufficient. For larger/more complex projects, multiple breakdown structures are likely to be used in order to define scope sufficiently.

The **product breakdown structure (PBS)** is a hierarchy of products (another term for deliverables or project outputs) required to be produced to complete the project. A

---

[1]   British Standards Institute (2019) "BS 6079:2019 Project management – Principles and guidance for the management of projects". British Standards Institute.

Code of Practice for Project Management for the Built Environment, Sixth Edition. Chartered Institute of Building.
© 2022 John Wiley & Sons Ltd. Published 2022 by John Wiley & Sons Ltd.

PBS uses nouns (names of deliverables at various levels). It is usual convention for the whole project to be referred to as Level 0 and then each subsequent breakdown of the project to be referred to as Level 1, Level 2, Level 3 etc. The lowest level of a PBS is a distinct project output such as a boiler or a balcony.

An example of PBS could include the headline processes of constructing a steel frame. This would include as a guide:

- dig footings for foundations

- build frame

- lay decking

- seal decking

The **work breakdown structure (WBS)** is a hierarchy of the activities to be done to complete the project. A WBS uses verbs and nouns. The lowest level of a WBS is either called a work package or alternatively an activity such as commission boiler or tile balcony.

One commonly used approach to define project scope is to start with a PBS to identify the main products, then at a certain level of definition, further break down those products into packages of work to be assigned to project team members to perform. This combined *technique is a useful way of capturing all the* work to be done.

The granularity of the lowest level of definition can vary but should be related to the techniques to be used to monitor progress. Best practice is to use Earned Value Management to monitor progress, and in this case, it is advisable for activities/work packages to be small enough that they do not span more than two reporting periods, for example two months if reporting is monthly, two weeks if reporting is weekly. At the first reporting period, the work is either not started, started or complete, avoiding overly precise and often incorrect estimates of progress.

Whatever technique is used (PBS, WBS or combined PBS/WBS), the objective is to define the scope of the project completely and in sufficient detail, so team members can do the work. It is particularly important when considering if ALL the work has been defined to include all the generic project management work and products (such as plans), as well as all the specific technical work and specialist outputs to the required quality and regulatory standards.

Irrespective of the technique used, each 'box' in the breakdown structure should be uniquely identified with a code.

An example of WBS could include the material processes of constructing a steel frame. This would include as a guide:

- concrete reinforcement and formwork for foundations

- structural steel frame bolts, fixtures and fittings

- decking materials, bolts, fixtures and fittings

- sealing materials and installation

Note: it is also common to show the project as a hierarchy of cost elements – the cost breakdown structure (CBS). This can be completed once estimates for each element of the PBS/OBS have been completed.

Responsible parties

The ***organisational breakdown structure (OBS)*** also represents the work of the project organised as a hierarchical breakdown of the management groups and resources involved in the project. An OBS often resembles an organisation chart and shows the project organisation in enough detail for work to be allocated to groups, units or to individuals.

Creating an OBS also allows creation of a responsibility matrix (often called a RACI chart) mapping the activities from a defined level of the WBS to the OBS. This enables a clear view of who is involved with the work and who is Responsible, Accountable, Consulted and Informed.

Quality planning

Adopting the definition in the International Standard for Quality Management ISO 9001:2015,[2] quality is the degree to which a set of inherent characteristics of an object fulfils requirements and where 'requirements' is defined as needs or expectations that are stated, generally implied, or obligatory.

Quality is related to the physical assets created and to the project management process. The purpose of quality planning is to enable the work to be done 'right first time' – preventing waste through re-work or rejected items.

For each element of scope, the quality requirements must be documented as acceptance criteria.

This information is often included in a quality plan, part of the integrated project execution plan.

Acceptance criteria need to be specific and observable/measurable and the plan needs to outline:

- The methods of verifying the outputs meet requirements

- Any pass/fail criteria for each method

- Frequency of any testing, checks or audits to be carried out

- Sample sizes for testing, and the resources and equipment needed to perform the test. These will often be provided from specialist consultancies if there is insufficient in-house capability or capacity.

- Who can approve results.

In some projects, all quality requirements will be in the same plan. In other projects, the requirements for regulatory compliance may be part of a related plan, e.g. a specific health and safety plan. The acceptance criteria for all scope items must be defined and included in the integrated *project execution plan*.

More information on quality planning and management can be accessed through the CIOB Guide.[3]

---

[2] International Standards Organization (2015) "ISO 9001:2015, Quality Management Systems – Requirements". International Standards Organization.

[3] Chartered Institute of Building (2021) "Guide to Construction Quality (Site Production and Assembly)". Chartered Institute of Building.

Managing scope and quality through the project

The version of the scope and quality plan in the PEP at the end of the Define phase will be the best version possible ahead of the Design activity. The PEP is agreed at the end of the Define phase to confirm it is sufficiently complete to start Design, but will continue to be updated under strict change control through Design.

Projects exist in a dynamic environment and there may be justified reasons to change scope and quality after the end of Design, but this should be by strict exception, and only with the authority of client governance. The concept of a 'change freeze' after the final investment decision at the end of the Design stage is common in many client organisations. See Guidance Note 42 on change control.

Guidance Note 30 Time planning and management

In the Identify and Assess stages of the project life cycle, high-level, top-down or comparative estimates of time need to be made but as noted in Guidance Note 14, the level of definition of estimates in early life cycle is very partial, and therefore the accuracy or maturity of estimates is low.

Once detailed planning of scope and quality is complete in the Define stage, detailed (bottom-up) time and resource planning can commence.[1]

The client project manager remains responsible for the adequacy and completeness of all plans for the project despite some of the detailed planning being completed by consultants and/or contractors in most scenarios.

Time planning can use a number of techniques to develop and present schedules (often called programmes in parts of the built environment) – a visual depiction of the work to be completed on a timeline.

This guidance note will refer to time plan as a generic term that encompasses what others may call schedules or programmes.

The standard and most-commonly used technique is critical path analysis (CPA).

In CPA, terms include network diagram (sometimes called a precedence network), three-point estimates, critical path, total float, free float, Gantt (bar) chart, milestones and baseline. The concepts are essential and it's important to understand them even if computer-based tools are used to create time plans.

Network diagram/Precedence network

This technique puts the activities defined in scope planning into a logical order. Logical dependencies need to be determined, i.e. predecessor and successor activities. Typical relationships used are:

- Finish to start (activity A must finish before activity B starts)
- Start to start (activity B can start when activity A has started)

1 British Standards Institute (2019) "BS 6079:2019 Project management – Principles and guidance for the management of projects". British Standards Institute.

Code of Practice for Project Management for the Built Environment, Sixth Edition. Chartered Institute of Building.
© 2022 John Wiley & Sons Ltd. Published 2022 by John Wiley & Sons Ltd.

• Finish to finish (activity B can finish when activity A has finished).

In considering dependencies, the time planner may also use the concept of 'leads' and 'lags' as shown in the figure, for example Activity B can start two days before Activity A has finished (Finish-Start dependency with a two-day lead), or Activity B can only start three days after Activity A has finished (Finish-Start dependency with a three-days lag).

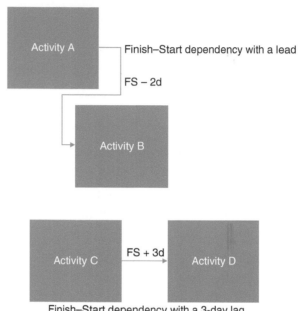

Time planning using 'leads' and 'flags'.

It is important to only model mandatory constraints at this point in the development of the time plan. Resource constraints will be added, where they exist, at a later stage. To consider them now adds a discretionary constraint that fixes an activity in time for a reason that could be managed.

Three-point estimates

Each activity in the network then needs an estimate of the time it will take to complete. Best practice is to use three-point estimates to reflect the uncertainty and risk in the activity (see also Guidance Notes 33–35). These are required if the project will perform quantitative risk analysis (schedule risk analysis alone or integrated cost and schedule risk analysis).

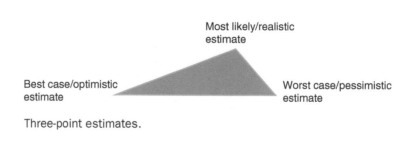

Three-point estimates.

The most likely estimate is typically skewed to the worst case as there are many more things that could go wrong, than go well with the activity.

The three points can be estimated independently (best educated guesses), or a more analytical technique can be used considering individual risks that apply to the activity.

Some planners will use the Program Evaluation and Review Technique (PERT) formula to calculate a single-point estimate to use in time planning using the established formula below. Common software also uses this formula to create single points from three-point estimates.

$$\frac{Best\,case + (4 \times Most\,likely) + Worst\,case}{6}$$

Critical path, total float and free float

The network diagram, with estimates for each activity, enables the critical path for the project to be determined.

All projects have at least one critical path. A critical path determines the shortest time in which a project can be completed by identifying the longest chain of activities through the project. Occasionally a project may have two chains of activity that are both equally critical.

The activities that are not on the critical path have float (alternatively known as slack). Total float is the time by which an activity may be delayed or extended without affecting the total project duration, whereas free float is the time by which an activity may be delayed or extended without affecting the start of any succeeding activity.

A project manager who understands which activities on their project have float, which are critical will be better able to make decisions relating to whether the start of an activity can be delayed or not.

Gantt chart

The most common way of showing a project's time line is to use a **Gantt chart** rather than the underlying network diagram. A Gantt chart (named after its creator) is a list of activities drawn against a horizontal timescale, with each activity represented by a bar depicting the planned duration.

Once the time plan has been fully developed and included in the PEP, it is said to be 'baselined'. The baseline is the version of the time plan against which the project time objectives will be monitored and controlled.

Milestones

In preparing project schedules, another key technique is the use of **milestones**. As the name suggests, milestones indicate key points in the project such as the completion of deliverables or decision points on the project. They are not activities because they have no (zero) duration.

Milestones can simplify the communication of the schedule by reporting the status of the project at a summary level. This kind of communication is essential for senior management or other parties who may not necessarily be interested in the detail of the project, but are interested in its outcome and progress. Milestones are also important for structuring payments to suppliers or contractors in some circumstances.

The unresourced time plan is a necessary starting point, then resource requirements and constraints are applied as described in Guidance Note 31.

Guidance Note **30**

Guidance Note 31 Resource planning and management

Resources are the people, equipment, facilities or any other entity that is needed to complete an activity and therefore cost money. Some resources are re-usable (such as people) and others are consumables (such as concrete). Rarely are resources in unlimited supply.

Most projects need to move from their ideal time plan – the one where all the work can be done to the quality required (see Guidance Note 30) to a version that is 'realistic' and takes into account resource availability. This is generically called resource management or resource optimisation. The concepts of **resource smoothing** and **resource levelling** are important because these are the ways that the project manager makes sure the work to be done is delivered at the best possible time given the project objectives and resources available.

Resource smoothing, also known as time-limited scheduling, is the process of making sure resources are used as efficiently as possible and of increasing or decreasing resources as required to protect the end date of the project. Using this method, time is relatively more important than cost.

Resource levelling, also known as resource-limited scheduling, is the process of making the most of the limited resources available. Resource levelling forces the amount of work scheduled not to exceed the limits of the resources available. This inevitably results in either activity durations being extended or entire activities being delayed until resource is available. Often this means a longer overall project duration. Using this method, cost is relatively more important than time.

One practical way of putting together the first iteration of a project timeline is to ignore the number or quantity of resources required, and the availability of these resources. The activities are sequenced and linked together graphically either using a network diagram and Gantt chart and then resources are considered as a second step to complete the picture.

In order to understand the complete picture, the resource or resources required to work on each activity must be linked to the schedule. This might be a person's name or a skill, e.g. designer, lawyer, or a company name, or a piece of equipment, such as a crane or a test rig. The level of detail and types of resources will depend on which party is preparing the time plan – the client, a consultant, the main contractor etc. The resourced time plan often displayed as a resource histogram. This is a graphical display of planned and/or actual resource usage over a period of time typically in the form of a vertical bar chart,

Code of Practice for Project Management for the Built Environment, Sixth Edition. Chartered Institute of Building.
© 2022 John Wiley & Sons Ltd. Published 2022 by John Wiley & Sons Ltd.

Guidance Note **31**

the height of each bar representing the quantity of resource usage in a given time unit.

Resources (and time) are often finite, and problems arise when the resources required exceed the resources that are available to the project. In such circumstances, two options are possible.

1. Re-schedule such that the need for additional resources is removed, but the end date of the project is maintained. This is known as time-limited scheduling or **resource smoothing**. This should always be the first thing that the project manager tries, but it doesn't always work if the activities that need to be moved are either on the critical path or have insufficient total float, the result being that the peaks either remain unchanged or are only reduced slightly.

2. Having unsuccessfully tried time-limited scheduling, the only option left is to address the peak in the most effective and least disruptive manner. This technique is known as resource-limited scheduling or **resource levelling** and typically leads to the project having a longer duration.

Resource critical path/Critical chain

An alternative technique, building on the work of Eli Goldratt and the theory of constraints, is a technique called (interchangeably) critical chain, or resource critical path. Critical chain works on the assumption most people tend to start a task at the latest start time not the earliest (student syndrome) and because many people are multi-tasking, they are not efficient in getting work done because the 'down-time' to switch between multiple tasks wastes time. Critical chain also addresses the fact that finishing a task late tends to be culturally unacceptable in organisations so hidden contingency or 'safety' is built into tasks to avoid being late. This time contingency is then often squandered by starting the task at the latest, not earliest, start time.

The alternative approach requires:

1. Building the schedule as normal, then

2. Halving the length of each activity (50% of the most likely), or using the optimistic estimate from the three-point estimate, and building the time taken out into a buffer to protect each critical chain of activity.

3. Allocating resources so they are not multi-tasking and should start the work at the earliest start time and hand-over as quickly as possible. Some tasks will use up some of the buffer, some will not – either needs to be satisfactory.

Not all projects are time critical, but where time is the driver, organisations using this approach report spectacular results. However, the critical chain approach relies on a culture being created within the project where it is accepted, the best-case estimates will rarely be achieved but on the understanding all the work will be completed as soon as possible and the buffer for the chain is there to protect the whole. Using this technique, main contractors and consultants need to be part of the planning and process in order to achieve the objective of completing the work as quickly as possible.

If resources were unlimited and always available, the critical path and critical chain methods would give the same result.

Other techniques that are sometimes used if time is critical are crashing and fast-tracking.

Crashing involves reducing the estimates of effort down to their minimum level (the best case) rather than the most likely or weighted average from the PERT formula. This creates a best-case schedule that is used to incentivise the team. Delays are inevitable, but are dealt with on a case-by-case basis rather than embedding what can be seen as 'padding' into the base schedule.

Fast-tracking takes different risks. In fast-tracking, the estimates of effort are not changed but the logical dependencies between activities are. An example might be work on one part of the scope is started ahead of a dependent part of the scope being finished. This can save time if all goes well, but results in re-work if the calculated risk back-fires.

All scheduling and resource management activities, however, rely on a complete definition of scope, a good understanding of logical dependencies and informed estimates of effort from the available resources.

More information on managing time in complex projects can be found in the CIOB Guide of that name.[1]

---

[1] Chartered Institute of Building (2011) "Guide to Good Practice in the Management of Time in Complex Projects". Wiley.

Guidance Note 32 Cost and budget planning and management

In the early part of the project life cycle, high-level, top-down or comparative estimates of cost need to be made but as noted in Guidance Note 14, the level of definition of estimates in early life cycle is very partial, and therefore the accuracy or maturity of estimates is low.

Once detailed planning of scope, quality, time and resource is complete, detailed cost plans and budgets can be created so that it is understood where capital and operational costs will be incurred over the life of the asset.[1]

The client project manager remains responsible for the adequacy and completeness of all plans for the project despite some of the detailed planning being completed by consultants and/or contractors in most scenarios.

Whereas the capital budget for creation of the asset should be defined in some detail at this stage (see Guidance Note 14 – classes of estimate), the operational costs needed to make up the whole-life cost are likely to be top-down or comparative estimates until the Design stage has been completed.

Knowing the likely out-turn cost of a project is important, but understanding how the costs will be incurred over time is vital to manage resource demand, supplier payments and cash-flow.

In some organisations, all resources that consume costs are included in the time plan, for example the volume of materials to be used. Other organisations only schedule labour and enter non-labour costs directly into a cost model.

Some organisations will create a cost breakdown structure (CBS), a hierarchical expression of the whole scope of the project but organised by cost items.

Where subcontracted resources are provided and committed contractually to provide as much labour or other resource as necessary to meet the timescales, the budgeted amount is the fixed price of the contract and remains at this level despite the actual resource used by the supplier.

The budget and cost plan is created by bringing together labour and non-labour costs. The cost profile for the resourced optimised time plan is known as the planned value (PV), or the budgeted cost of work scheduled (BCWS). The BCWS is the cost profile against which the project measures progress (see Guidance Note 37 – monitoring, measuring and reporting progress).

---

[1] British Standards Institute (2019) "BS 6079:2019 Project management – Principles and guidance for the management of projects". British Standards Institute.

Code of Practice for Project Management for the Built Environment, Sixth Edition. Chartered Institute of Building.
© 2022 John Wiley & Sons Ltd. Published 2022 by John Wiley & Sons Ltd.

Most organisations will also have financial guidance for how to deal with exchange rates, inflation and other variables that could change the actual cost of the project but are out of the project manager's control. Exceptions could be SMEs where the client project manager is working directly to the senior leadership team and takes responsibility for all influences on project and whole-life costs.

The final point to make about budgeting is about cost contingency. It is usual for organisations to include a contingency to cover unforeseen costs in their estimated cost. The simplest and least accurate way of doing this is to add a percentage depending on the perceived risk in the plan. The most thorough way of doing this is to build up the contingency figure by looking at the combined effects of estimating uncertainty and the risks identified in the risk log. This is described in Guidance Notes 32 and 33.

More details on how to deal with aspects of cost planning and budget such as how to deal with fixed and variable costs, or how to deal with allowances or other contractual payments are covered in the UK government cost estimating guidance.[2]

2   UK Cabinet Office – Infrastructure Projects Authority. Cost Estimating Guidance. Available at https://assets.publishing. service.gov.uk/government/uploads/system/uploads/attachment_data/file/970022/IPA_Cost_Estimating_Guidance.pdf (accessed 18 March 2021).

Guidance Note **32**

Guidance Note 33   Risk identification

Note: this publication uses terminology as specified in the International Standard for Risk Management[1] and the associated International Guide to Risk Vocabulary[2]. Your organisation may use other interchangeable terms, for example risk assessment instead of risk analysis, or risk responses instead of risk treatments.

Risk identification is a continuous process undertaken in a structured way at different points of the project life cycle and at different levels of the project/programme/portfolio structure.

According to the International Standard for Risk Management, risk is the '**effect of uncertainty on objectives**'. In Guidance Note 2 on risk appetite and delegated limits of authority, the need to define the appetite for risk in measurable terms and to use this to calibrate impact scales to be used in risk analysis and evaluation was outlined. This work is necessary to make the project objectives clear and measurable and to make it clear when a risk needs to be escalated, or delegated.

Risks may be threats (downside impact on objectives) or opportunities (upside impact on objectives).

With clear and measurable objectives in place, risks (uncertainties that would effect those objectives) can be identified. It is vital when identifying risks to make it clear exactly what is at risk and why.

To enable effective analysis and evaluation, risk descriptions are needed to clearly separate what is known (a fact or issue now, or an assumption held as true for planning purposes) from what is uncertain (may or might or could happen) from the objectives at risk.

Commonly used ways to do this are shown below. It is the same risk described in two different ways and either is acceptable.

Code of Practice for Project Management for the Built Environment, Sixth Edition. Chartered Institute of Building.
© 2022 John Wiley & Sons Ltd. Published 2022 by John Wiley & Sons Ltd.

SOURCE (FACTS NOW)	RISK (UNCERTAIN EVENT)	EFFECT (ON OBJECTIVES)
Because the design will use materials that are unfamiliar to us	we may experience unforeseeable challenges in construction leading to	delays and cost over-runs.

RISK (UNCERTAIN EVENT)	EFFECT (ON OBJECTIVES)	SOURCE (FACTS NOW)
There is a risk that we may experience unforeseeable challenges in construction	leading to delays and cost over-runs	because the design uses materials that are unfamiliar to us.

It is best practice to avoid shorter risk descriptions that do not clearly separate what is known/fact now from what is unknown/uncertain. Risk descriptions should be clear enough, so a non-risk expert is able to understand them, and so risk owners can estimate likelihood of occurrence and magnitude of impact.

Avoid risk descriptions with multiple uncertainties, if x occurs, y may occur.

Engaging relevant stakeholders in risk identification is important. Avoid limiting their input to the things that you agree with. Risk identification, done well, is a creative and divergent exercise that describes all the uncertainties that would effect objectives, if they occurred.

There are many techniques available to support risk identification including:

- **Checklists** based on previous experience

- **Prompt lists**, for example using PESTLE (political, economic, sociological, technological, legal, environment), or an alternative that suits the project

- **Interviews**, or other form of individual input, for example the Delphi technique where people are not influenced by others

- **Brainstorming**, or other form of group session where people can 'bounce off' the ideas and perspectives of others

- **Structured approach,** for example a SWOT analysis to identify organisational strengths and weaknesses and project threats and opportunities that relate to these.

Identified risks need to be given a unique identifier and, ideally, stored in a single repository for risk information for the project. Where the project does not have a single information system for risks, or where contractual conditions do not motivate sharing of risk information, and so there are multiple, separate risk registers; the client project manager needs to devise clear rules for sharing, escalation and delegation, so there is a greater chance of achieving project objectives.

Once the risks perceived by stakeholders are clearly identified and described, they can be analysed to prioritise their importance and guide investments in risk treatments. See Guidance Note 34, risk analysis and evaluation; Guidance Note 35 quantitative risk evaluation; and Guidance Note 38, risk treatment.

Guidance Note 34   Risk analysis and evaluation

Note: this publication uses terminology as specified in the International Standard for Risk Management[1] and the associated International Guide to Risk Vocabulary.[2] Your organisation may use other interchangeable terms, for example risk assessment instead of risk analysis and evaluation, or risk responses instead of risk treatments.

Risk analysis and evaluation (and subsequent risk treatment) happens at different times and at different levels across the project life cycle, for example within a contractor's organisation, within a single stage, or overall.

Risk analysis is a prioritisation process, to decide which risks warrant the most urgent attention to increase certainty of achieving project objectives or to prepare contingency plans. Risk evaluation is a decision process, what investments will be made in increasing certainty?

Quantitative risk analysis, looking at the combined effects of estimating uncertainty and specific risk events is covered in Guidance Note 35.

In qualitative risk analysis, it is necessary to estimate two parameters:

- The likelihood of the risk occurring
- The magnitude of the consequence on objectives

Likelihood of occurrence

If the risk was a game of pure chance like playing the lottery, or where you have a large homogeneous set of data that is relevant, you could calculate the probability of the risk occurring. In projects this is rarely, if ever, possible, and so you need to estimate the likelihood of occurrence. There are different methods used to do this, some using past frequency of risks materialising as a proxy for future likelihood. No method is accurate (we are predicting the future!). False precision is futile.

Most organisations use a five-point scale (although a three-point scale is also acceptable for a less granular prioritisation). Typical useful scales that can be used for likelihood include:

---

[1] International Organization for Standardization (2018) *ISO 31000:2018 Risk Management – Guidance*. International Organization for Standardization.

[2] International Organization for Standardization (2009) *ISO Guide 73 Risk Management – Vocabulary*. International Organization for Standardization.

Code of Practice for Project Management for the Built Environment, Sixth Edition. Chartered Institute of Building.
© 2022 John Wiley & Sons Ltd. Published 2022 by John Wiley & Sons Ltd.

Likelihood of risk occurrence: 5-point scale

Likelihood score	Descriptors	Percentages	Frequency (based on past experience)
5	Almost certain	>80%	Happens multiple times in every project
4	More likely than not	50–80%	Happens once in every project
3	Equally likely or unlikely	50/50	Happens in half of all projects
2	Less likely than not	20–50%	Happens in a few projects
1	Unlikely	<20%	Never known it to happen

Magnitude of consequence (size of impact)

Consequence or impact scales are derived from the work done to identify objectives at risk and the appetite for risk to those objectives (see Guidance Note 2). It is vital that the narrative on effects of risks align with the selected consequence scale.

Most organisations use a five-point scale (although a three-point scale is also acceptable for a less granular prioritisation).

There should be a scale for each relevant objective. The highest magnitude should represent a catastrophic size of consequence/impact – a level that is out of appetite for the project. The lower magnitude should represent an insignificant size of consequence/impact for the project. Best practice is to use a halving technique for the levels between, for example:

Likelihood and financial impact of risk occurrence: 5-point scale

Consequence score	Cash (capex or opex)	Time to completion	Safety
5	>£40,000	>26 weeks	Fatality
4	£20,000–£40,000	13–26 months	Permanent disability
3	£10,000–£20,000	7–13 weeks	Reportable lost time accident
2	£5000–£10,000	3–7 weeks	Non-reportable lost time accident
1	<£5000	<3 weeks	Minor injury or near miss

Analysing risks for likelihood and magnitude of consequence enables the first level of prioritisation.

Some organisations use a second-order sorting process considering one of the following:

Proximity: a time-related factor, e.g. when would the risk occur if it did

Velocity: how quickly would the consequence happen if the risk occurred

Urgency: the time window when action might be possible

It is also increasingly popular (and software exists) to consider the network of risks, looking at connections between risks to identify, systemically, which risks would have the greater impact should they occur as shown in the figure below.

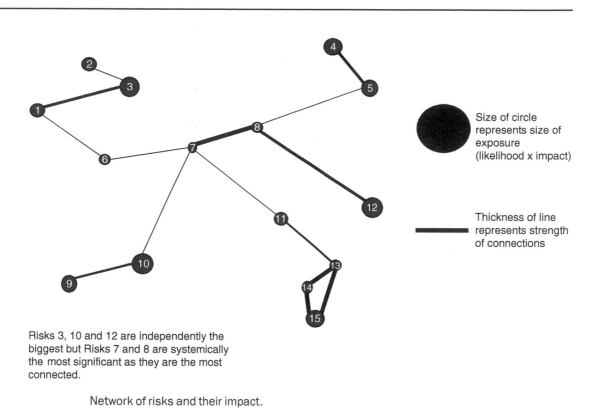

Size of circle represents size of exposure (likelihood x impact)

Thickness of line represents strength of connections

Risks 3, 10 and 12 are independently the biggest but Risks 7 and 8 are systemically the most significant as they are the most connected.

Network of risks and their impact.

Consistent scoring

A common mistake with risk analysis is the failure to normalise the thinking of the people who estimate likelihood and consequence.

The most normal practice in projects is to analyse the current or 'net' exposure of each risk.

This is the likelihood and magnitude of consequence of each risk assuming all normal controls are implemented and effective. The alternative is to analyse the inherent or 'gross' exposure of each risk, assuming no controls are in place, or all controls fail.

Take, for example, the likelihood of the risk that procurement staff would accept bribes from contractors. The inherent likelihood of this occurring is probably higher than the current likelihood because it would be normal for procurement staff to be well versed in policy and legislation and motivated to comply.

Some organisations use a 'bow-tie' visualisation of risks to highlight preventive controls and corrective controls, and this can be useful in ensuring that reasonable judgements are made about the effectiveness of existing controls, prior to any further risk treatment.

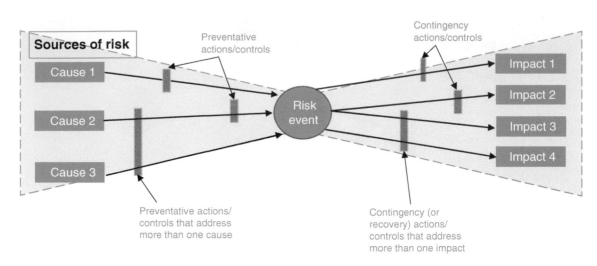

Bow-tie of risks and preventative controls.

Once the risks perceived by stakeholders are analysed to prioritise their importance, risks can be evaluated and decisions made to guide investments in risk treatments or other actions, e.g. to focus on ensuring the effectiveness of existing controls.

In addition to qualitative risk analysis as described here, many projects use quantitative risk analysis to look at the combined effects of risks on the project business case, budget or time-plan. This is described in Guidance Note 35. Risk treatment is covered in Guidance Note 38.

Guidance Note 35   Quantitative risk analysis and evaluation

Qualitative risk analysis and evaluation, as described in Guidance Note 34, is focused on the prioritisation of individual risks.

Quantitative risk analysis and evaluation is focused on determining the combined effects of all estimating uncertainty and risk events on the project's business case, whole-life cost, capital budget and/or timeline. This is known by different terminology in different companies, including:

- Cost risk analysis (CRA)

- Schedule risk analysis (SRA)

- Cost and schedule risk analysis (CSRA)[1]

- Probabilistic risk analysis (PRA), or just

- Quantitative risk analysis (QRA)

Some organisations, not all, add Q to all the acronyms, e.g. quantitative cost risk analysis (QCRA).

This type of analysis is important to:

- Consider uncertain/variable parameters such as productivity rates, the cost of steel or weather patterns alongside discrete risk events.

- Provide further insight into which risks warrant proactive responses.

- Understand different potential scenarios/stress test situations.

- Manage stakeholder expectations about confidence levels in achieving particular out-turn times, costs, return on investment.

- Size financial contingency in a more sophisticated way than crudely adding a percentage of budget (for more information on contingency planning and management, see Guidance Note 36).

- Monitor trends of levels of risk exposure vs. risk contingency.

---

[1] **Integrated Cost-Schedule Risk Analysis** by David T. Hulett and Michael R. Nosbisch, CCC PSP, *Cost Engineering*, Vol. 54, No. 06, AACE International, Morgantown, WV, 2012. *Figure 4 and Figure 6* used with permission of AACE International, 726 East Park Ave., #180, Fairmont, WV 26554. Email: info@aacei.org; Phone 304.296.8444; website: web.aacei.org.

The most common method is to build a model that maps all estimating uncertainty and risk events to the costs/timeline/IRR (whatever parameter is the focus of the analysis). Project organisations will typically have licenses for proprietary software to do this work.

Once the model has been validated to ensure that it is complete and the logic is sound, a random number generator (typically Monte Carlo simulation) is performed to build a view of the combined effects of all variables.

Validation of the model is vital to prevent errors that would render the outputs unreliable, for example where risks would affect more than one element of the scope, the model reflects this. A risk that ground conditions might be worse than assumed in the base estimate would affect all line items in the timeline or cost model relating to that ground. The ground conditions would not be as estimated in one place, better than expected in another and worse than expected in another. Correlating related items is one vital contributor to a reliable output.

This Guidance Note cannot provide a detailed 'how-to' guide for practitioners. Creating robust risk models is specialist work, with many pitfalls. Client organisations will often procure specialist resources from consultancies to assist with this work.

Communicating outputs to decision-makers to enable evaluation

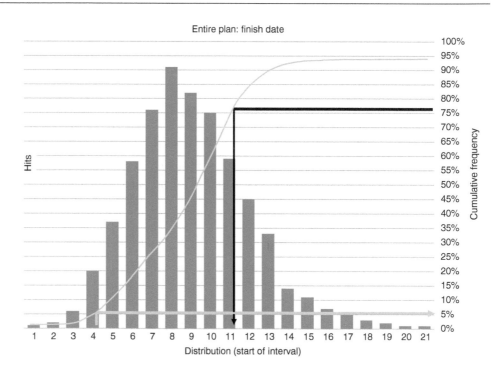

Monte Carlo diagram showing confidence levels in achieving 'un-risked' plans. Source: Integrated Cost-Schedule Risk Analysis by David T. Hulett and Michael R. Nosbisch, CCC PSP. Cost Engineering, Vol. 54, No. 06, AACE International, Morgantown, WV, 2012. © 2012 AACE International, all rights reserved.

Outputs from a quantitative risk analysis enables decision-makers to consider confidence levels in achieving 'un-risked' (deterministic) plans.

In the example shown, the analysis shows that there is a 5% chance of achieving the original planned finish date of <define> and that decision-makers should have 50% confidence of completion by <define date> – the P50, and 80% confidence of completion by <define date> – the P80.

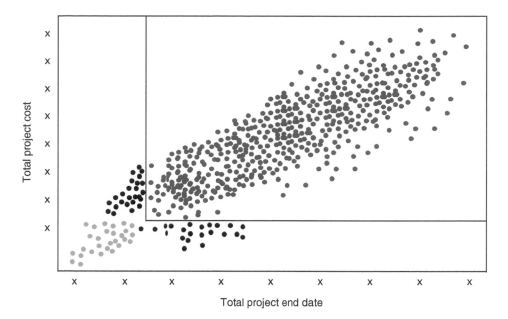

Scattergram showing incidences from the simulation that the project would be late and over-budget. Source: Integrated Cost-Schedule Risk Analysis by David T. Hulett and Michael R. Nosbisch, CCC PSP. Cost Engineering, Vol. 54, No. 06, AACE International, Morgantown, WV, 2012. © 2012 AACE International, all rights reserved.

How this analysis can be used to plan and manage contingency is covered in Guidance Note 36.

Other useful outputs include plots that show time and cost on one graph, for example:

The points plotted show the individual simulated outputs that have the project out-turns ahead of or behind schedule, and below or above budget. The red

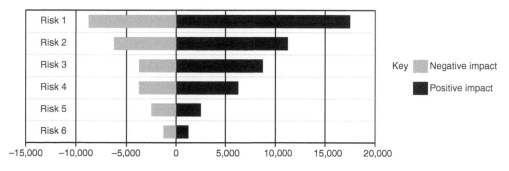

Toronto diagram showing the risks with the largest overall impact on objectives.

dots represent the incidences from the simulation that the project would be late and over-budget. In the picture, there is a 95% chance of overrunning both time and cost, and the model shows that time and cost are 77% correlated. Not all projects are so optimistically planned, but some are, and this analysis helps the client, funders and other key stakeholders to have realistic expectations.

Outputs can also show the risks that have the largest overall impact on objectives, as shown below.

This is a good reminder that risks are defined as 'the effect of uncertainty on objectives'. Risks can have a negative impact or a positive impact on objectives. Quantitative risk analysis enables this to be modelled and for decisions to be made with far better data available than a qualitative risk analysis considering only likelihood and magnitude of consequence of risk events in isolation.

All of this analysis is designed to help to prioritise focus and increase chances of meeting project objectives.

Guidance Note 36   Contingency planning and management

Once plans have been made reflecting scope, quality, time, resources and costs, and in the light of the estimating uncertainty and specific risks to project objectives, the client needs to decide how much contingency to hold. The level and degree of contingency planning needs to be compared and contrasted to the actual project risks. This analysis should take place in the pre-construction phase and then be monitored throughout the construction phase until project completion.

Contingency, typically expressed in financial terms, is set aside to respond to identified risks. It is needed to match the gap between the 'un-risked' plan and the desired level of confidence that decision-makers require.

In addition to a contingency for identified risks – sometimes known as the risk budget – organisations will hold a management reserve to make provision for unidentified (ideally unknowable) risks or those risks that have a very low likelihood of occurrence but would have a very high impact if they did.

Different terms are used by different organisations, but in simple terms this is how financial contingency is made up. For a more detailed view, the UK government cost-estimating guidance has useful information.[1]

Different organisations will use different methods to determine contingency, and this is a feature of the maturity of project management in that organisation and the scale/complexity of the project.

If it is normal in the organisation to use quantitative risk analysis as described in Guidance Note 35, then confidence levels can be used to size contingency. For example, the allocated provision (un-risked budget) could be set at the P50 (50%) confidence level, with a risk budget to the P70 and management reserve to the P90. Much depends on whether the investment in the project is a one-off for the organisation, or whether it is part of a portfolio of projects. Other determinants include the certainty of funding, for example it may be judged not possible to go back to funders for more money at a later date.

---

[1] UK Cabinet Office – Infrastructure Projects Authority. Cost Estimating Guidance. Available at https://assets.publishing. service.gov.uk/government/uploads/system/uploads/attachment_data/file/970022/IPA_Cost_Estimating_Guidance.pdf (accessed 18 March 2021).

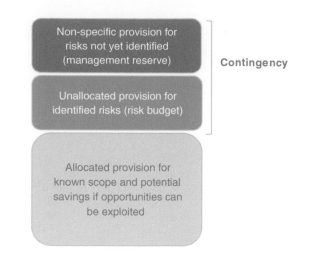

Financial contingency.

Other methods include aggregating expected monetary values (EMV) for risks in the risk register (likelihood x size of financial impact) or adding a percentage based on experience. Although no method is perfect, both of these methods are crude in comparison to using the outputs of quantitative risk analysis to size contingency and to manage contingency run-down.

It is particularly difficult to manage contingency run-down accurately using EMVs of individual risks or a percentage as the provision is based on a likelihood of the risk occurring, for example contingency provision for a risk with a 50% chance of occurrence but would cost £1m if it occurred would be £500K. Yet if the risk occurs, £1m is required, not £500K.

Best practice is to run quantitative, probabilistic analysis at each stage gate as a minimum and then re-size contingency accordingly.

Most organisations will proportion the allocation of contingency between the client project manager and the client sponsor and governance board. Some organisations hold the whole contingency at sponsor/governance board level. This reflects confidence in the planning process and the amount of control that is desired by the client. Contingency is for identified risks, not to spend on scope that was omitted from original plans.

Guidance for public investments in the United Kingdom is provided in the latest version of the Green Book.[2]

Guidance Note **36**

---

[2] HM Treasury (2020) "The Green Book: appraisal and evaluation in central government". Available at https://www.gov.uk/government/publications/the-green-book-appraisal-and-evaluation-in-central-governent (accessed 17 May 2021).

Guidance Note 37 Progress monitoring, measuring and reporting

Monitoring progress requires three things:

1. A baseline to monitor against

2. Data on actual performance

3. An assessment of the implications of performance on the rest of the plan

Combined with data on any changes in the external context, performance monitoring enables decision-makers on governance boards to take corrective action where necessary.

The project execution plan (PEP) provides the baseline for all aspects of the project's plans (Guidance Note 28) including:

- Achievement of scope to the right quality

- Achievement of timelines

- Use of resources – people, equipment and the associated costs

- Cashflows

- Performance of contractors

- Benefit realisation

- Stakeholder satisfaction

Reports are often provided using 'traffic lights' to indicate the performance to date and to forecast.

Best practice in tracking performance of time and cost is to use Earned Value Analysis (EVA). The baseline is the Budgeted Cost of Work Scheduled (BCWS) or Planned Value (PV) – see Guidance Note 12.

Progress is measured by comparing the work achieved (earned) in the time and the work achieved (earned) for the money spent. A straight comparison of actual spend vs budget gives a partial picture – it does not indicate the work for the money in the time. Using EVA, the schedule and cost variances can be calculated and used to forecast the time and cost at completion assuming no corrective action was possible and/or implemented.

Code of Practice for Project Management for the Built Environment, Sixth Edition. Chartered Institute of Building.
© 2022 John Wiley & Sons Ltd. Published 2022 by John Wiley & Sons Ltd.

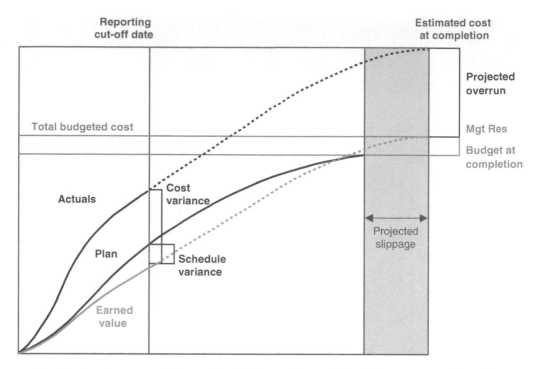

Tracking performance of time and cost. Source: Earned Value Management – How to?, Pramod Rao, DIGITAL BUSINESS TRANSFORMATION + PROGRAM MANAGEMENT, March 12, 2011.

Guidance Note 38   Risk treatment

Risk treatment is the terminology specified in the International Standard for Risk Management[1] and the associated International Guide to Risk Vocabulary.[2] Your organisation may use other interchangeable terms, for example risk responses or risk mitigations.

When risks have been identified, analysed and evaluated (see Guidance Notes 33–35), risk treatments can be implemented for priority risks, i.e. those where investing in increasing certainty has a pay back for the project – is 'worth it'.

Treating priority risks may involve one or more of the following.

- Avoiding the risk by deciding not to start or continue with the activity that gives rise to the risk (terminate the risk)

- Pursuing the risk in order to achieve the goal (take the risk)

- Removing the risk source/cause (terminate the risk)

- Changing the likelihood (preventive control – a form of treating the risk)

- Changing the consequences (corrective control – a form of treating the risk)

- Sharing the risk e.g. through contracts or buying insurance (transfer the risk)

- Managing the consequences if risks materialise

Some risks will be low priority and will be tolerated, i.e. no action taken other than to track if the likelihood and/or consequence changes.

A 'bow-tie' visualisation of a risk (also shown in Guidance Note 34 – risk analysis and evaluation) will enable the team to focus on 'normal' controls – things that should be working effectively now, and on 'new' controls to further manage the risk. New controls may be preventive or corrective, i.e. including new 'plan B's'.

---

[1] International Organization for Standardization (2018) *ISO 31000:2018 Risk Management – Guidance*. International Organization for Standardization.

[2] International Organization for Standardization (2009) *ISO Guide 73 Risk Management – Vocabulary*. International Organization for Standardization.

Code of Practice for Project Management for the Built Environment, Sixth Edition. Chartered Institute of Building.
© 2022 John Wiley & Sons Ltd. Published 2022 by John Wiley & Sons Ltd.

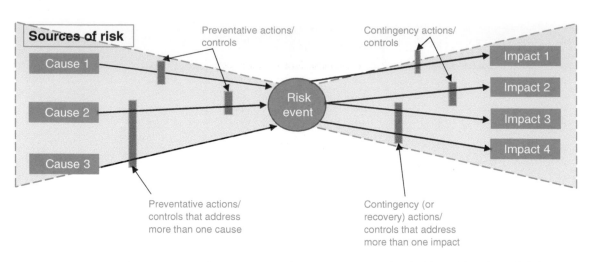

Investments in new controls must be 'worth it'. The objective of risk treatment is to plan and implement actions that bring the exposure of risks into a tolerable range based on the scales calibrated to represent the risk appetite of the client.

If the risk work is being undertaken by a contractor in the supply chain, calibration of impact scales for the contractor need to reflect the client's appetite, i.e. it is clear where knowledge of a risk being managed by the contractor should be shared with the client.

In the figure, the current and target positions for risks are plotted with a key to show the risk owner's assessment of the effectiveness of the controls designed to treat the risk.

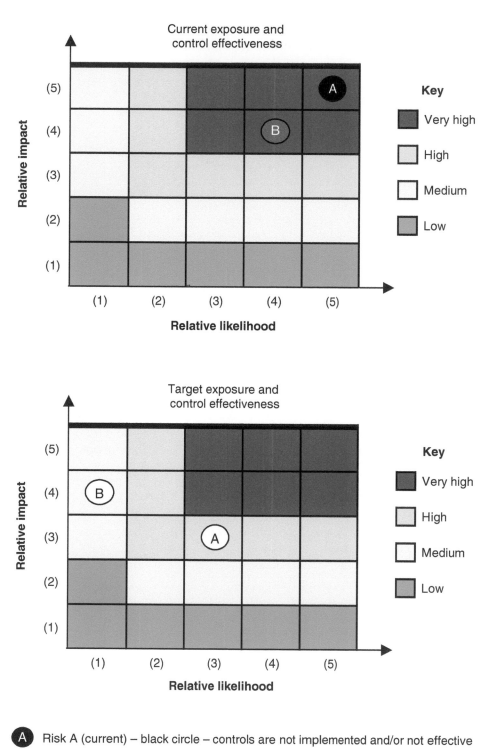

A Risk A (current) – black circle – controls are not implemented and/or not effective

(A) Risk A (target) – white circle – controls are fully implemented and effective

(B) Risk B (current) – blue circle – controls are not implemented and/or not effective

(B) Risk B (target) – white circle – controls are fully implemented and effective

Tabulation of current and target positions for risks, showing owner assessment of effectiveness of risk treatments.

If the project is using quantitative risk analysis, good practice is to compare the pre-treatment and post-treatment models so that the combined effect of risk treatments can be seen. This provides an objective way of assessing whether the investments of time and money in risk treatments have a significant enough effect on the project objectives.

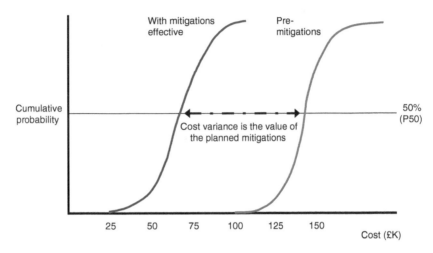

Comparison of pre-treatment and post-treatment models.

Plans to treat risks are just that. Planned work should be added to the baseline plans in the Project Execution Plan, i.e. it is new scope with quality requirements that will use time and resources and will cost the project.

The allocated risk owner is responsible for tracking progress and recommending to governance when the risk exposure is tolerable, or where further action is warranted as the exposure is increasing.

Guidance Note 39 Issue resolution and problem solving

Issue resolution and problem solving is an essential skill for those working in project management.

This guidance emphasises solid, staged steps to manage problems that may arise in the most effective way whilst minimising any detrimental impact to the project – thus providing a sound foundation for individuals to develop their own methodology, each appropriate to their own role and industry specialisation.

It is vital to note that where an issue or problem results in a change to the approved *project execution plan* and/or *business case*, the formal change control process needs to be followed – see Guidance Note 42.

All project managers should be aware that in the event of a problem, they should not lose sight of the primary tasks, which are to construct a project on time and on budget with the proper health and safety measures in place.

In the built environment, this could be as simple as making sure the site is running smoothly to enable either the Project Manager, the Management Team or simply a nominated individual within the team to then deal with the problem.

It should be noted, however, that construction sites allow additional resources to be brought in, if and when required – which can reduce pressure on the project manager, potentially speed up the resolution process and consequently control cost. However, due to the transient and nomadic nature of the construction industry, it is essential that the additional resources are fully trained and aware of the projects time, cost and health and safety requirements.

Situational awareness

The perception of the elements in the environment within a volume of time and space, the comprehension of their meaning and the projection of their status in the near future

Or more simply:

Knowing what is going on around you

Clearly situational awareness is a valuable trait, or skill in the Project Managers armoury, and conversely, failure by the Project Manager to maintain situational awareness when problems arise can easily allow a relatively minor, but

Code of Practice for Project Management for the Built Environment, Sixth Edition. Chartered Institute of Building.
© 2022 John Wiley & Sons Ltd. Published 2022 by John Wiley & Sons Ltd.

Guidance Note **39**

unexpected, problem to become a distraction and appear complex, which could then easily derail part of or even a whole project – in regard to any loss of time, productivity, programme, costs etc.

With this in mind, in the earliest stage of recognising a problem (which cannot be immediately resolved by the Project Manager), generally two options are presented:

1. To determine the complexity of the problem, then allocate responsibility for dealing with the issue to a specific person/party.

Or

2. Give responsibility/ownership of dealing with the problem to an individual, and then review the complexity.

The challenge therefore is in determining the most appropriate option, and the following steps will, in the vast majority of cases, assist:

'Yes', 'No' or Err. . .

As is often the case, there is no right or wrong in this process, as many factors (both on-site and off) may apply. Therefore, the quick and simple 'Yes, No, Err' approach can be applied when facing the unexpected:

Question: '*As the Project Manager can I deal with this problem relatively quickly, without undermining the main task?*'

- If the immediate answer is 'Yes', then the Project Manager should deal with the problem without delay, and then return to their primary tasks.

- However, if the answer is a resounding 'No', it is likely to be more appropriate to allocate the task to another person, who in turn may elect to bring in additional resources and/or specialist advice.

- However, more often than not, the response will be an uncertain 'Err. . .', and which almost certainly should be taken as a 'No' – as to delay the decision, or overburden the Project Manager is likely to compound the problem.

Hesitation and prevarication, along with 'knee-jerk' reactions can waste valuable time, and subsequently lead to further issues that may have been avoided, had a prompt informed decision been made in the first instance.

With this in mind, the following steps could prove to be invaluable:

Who's best placed to deal with the problem: It's human nature to believe that you need to own the problem that faces you, and as the Project Manager, you probably do – or at least should own it. However, despite your wealth of experience, you may not be the best individual to deal with the matter for a host of reasons; current workload, a technical issue outside your field of experience, or perhaps it's a relatively minor but key matter that needs to be resolved – however don't be afraid to delegate!

Delegation: The options are numerous and are dependent on many factors; the complexity of the issue, the need (or otherwise) to engage an SME on the matter in hand, the resources at hand, the resources available (if called in), or perhaps the issue is not so great, and perhaps it lends itself to a development opportunity to a junior team member?

Get the facts: Once appointed, the owner of the problem should then determine, and use the facts (as known at the time), and never rely on assumptions and/or beliefs of either their own or those of others – as valuable time, money and resources can easily be wasted.

Brainstorm: Where there are no obvious answers, the solution will probably come from someone in the wider team, or as a combined effort – construction is a team game, so do not be afraid to seek the input of others.

Consider the possible outcomes: As potential solutions emerge, consider the likely impact considering time, cost, impact on project etc., and make an informed decision OR advise others, so they can make that decision if it is not your remit.

Remain flexible: Even though the solution appears sound, and the wheels are in motion, the progress and/or improvement in the situation must be monitored. Some problems can be 'fluid' and changes in the situation could mean the urge to 'press on' with the original plan may no longer be the best option, and so a complete change of approach may be required.

Communicate: Once you have the facts, appraise those who may be impacted, detailing the issue at hand and the potential outcome(s), a brief overview of the options and the suggested resolution(s). And dependent upon the recipients, provide details of likely risks to the programme and/or cost implications.

Implement & Monitor: Once a plan is agreed, swift implementation and ongoing monitoring is essential, so that any corrective actions can be taken to 'fine tune' the work in progress, or if necessary to re-think the situation.

Guidance Note 40 Health and safety plan

The health and safety plan is a live document that should be initiated at the earliest possible opportunity. It should start with details of the organisation, the location and the description of the work. The plan should have a quality assurance process where it is clearly evidenced who prepared and who approved the plan.

A template of contents should include as a minimum, but not be limited to, the follow:

- The health and safety policy statement
- The schedule of appointed responsible persons
- The management and supervision organisational chart
- Risk assessments
- Fall protection plans
- Hazardous work activities – including method statements
- Personal protective equipment requirements
- Measures to control the condition and use of tools and equipment
- Fire protection, prevention and control measures
- The use of control of hazardous chemical substances
- Environmental protection measures
- First aid arrangements
- Construction site traffic management, logistics and signage

It is also recommended that the following appendices are included as part of the health and safety plan:

- Client health and safety specification
- Principal contractor appointment
- Appointment letters
- Personal protective equipment (PPE) issue registers

Code of Practice for Project Management for the Built Environment, Sixth Edition. Chartered Institute of Building.
© 2022 John Wiley & Sons Ltd. Published 2022 by John Wiley & Sons Ltd.

- Equipment, inspection and repair registers

- Material safety data sheets

- Safe disposal certificates

It is important that the health and safety plan is agreed to prior to works commencing on the project and updated at regular intervals.

Guidance Note 41 Preparation for stage gate reviews

The purpose of a stage gate is to review progress and confirm the viability of the work completed. In all cases, the client sponsor and wider governance board is accountable for decisions made – whether that is to continue as planned, continue with approved changes, or stop.

In each chapter of this *Code of Practice,* the decisions to be made at each stage are detailed.

This guidance note focuses therefore not on the decisions required, but on the preparation for taking those decisions.

It is perhaps understandable that project managers, in preparing for a stage gate, are primarily motivated to 'pass' the gate and have permission to continue. This is, however, the wrong focus – the primary responsibility of the project manager is to help the client to make the right decision about how to invest their funds. Preparation, therefore, is about gathering and presenting the best information possible to support decision-making.

Evidence gathered through monitoring, measuring and reporting processes at regular intervals during any particular stage will form the basis for preparation: the information that provides an accurate picture of progress and any variations from plan.

Depending on the governance arrangements in place (see Guidance Note 3 – assurance across the three lines of defence), decision makers will require assurance activities to have taken place to provide them with a view, independent from the people managing the work, about progress and opportunities for learning. This may include assurance activities involving external stakeholders, as well as members of the team, whether staff, consultants or contractors. Assurance activities may be formal third-party audits, or less formal assessments made by support functions such as a project management office (often referred to as project controls or project services).

Understanding progress and the reasons for deviations from plans is vital, but the core focus of preparation for stage gates should set the historic evidence in the context of the future. Preparation for stage-gates therefore must include:

- Up-to-date risk analysis and forecasts (see Guidance Notes 33–35).

- Any emerging data that changes the premise of the project, for example that would change end-user needs and expectations, where there is new

Code of Practice for Project Management for the Built Environment, Sixth Edition. Chartered Institute of Building.
© 2022 John Wiley & Sons Ltd. Published 2022 by John Wiley & Sons Ltd.

legislation or industry guidance or where macro-economic changes put the business case in question.

In many organisations, there will be formal requirements for preparation and provision of papers for a stage-gate review meeting (alternatively called a Steering Group, Working Group, Project Board and other variants). In such situations, the decisions that are required must be made clear, and information presented to enable that decision to be made.

Many organisations have developed a mature culture for decision-making where attendees will have reviewed papers in advance and will be ready to focus the meeting on discussion rather than the project manager using all the available time to present the papers with little time for meaningful challenge and debate. In less-mature decision-making cultures what is observed is a process of a few stakeholders taking decisions before the stage-gate meeting and using the meeting to 'rubber-stamp' the outcome. This is sub-optimal and may lead to missed opportunities to improve delivery, anticipate risks and safeguard the investment.

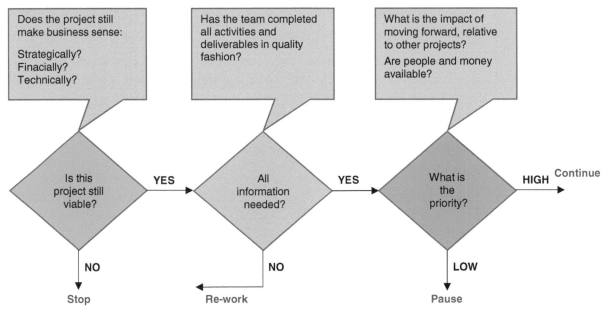

Less-mature decision-making process.

Guidance Note 42   Change control

Change control must be applied to all elements of the project once this is approved and baselined at the end of the Define stage of the life cycle. This includes the project information, including all elements of the project execution plan (PEP) and the version of the intermediate business plan that matches the PEP at the end of Define, and all elements of the asset information that exist at this time.

For change control to be effective, all discrete pieces of information, for example documents, drawings, models etc. must have a unique identifier. Each uniquely identified piece of information is known as a configuration item in some environments. The unique identifier also provides a means of having a clear map of related items so that a change to one piece of information triggers a review of related information. An example would be a quality plan, where a change to acceptance criteria for deliverables would trigger a review of the related time plan and budget.

Any changes that would affect contracts with suppliers in the supply chain are also within the scope of change control.

Note: In some environments, change control is used interchangeably with the term change management. In other environments, change management is a different discipline focused on planning and managing organisational change, not managing changes to project and asset information.

Changes may arise from many places, including:

- A change in justification for the project

- A change in requirements to the scope or specification of deliverables; perhaps due to a change in legislation or regulation

- The discovery of a mistake or error such that the delivered outputs will no longer work

- An incorrect estimate of time or money

- A reduction in resource availability

- A technological or market place change

Code of Practice for Project Management for the Built Environment, Sixth Edition. Chartered Institute of Building.
© 2022 John Wiley & Sons Ltd. Published 2022 by John Wiley & Sons Ltd.

Changes may come from identified risks that are now facts not uncertainties, or from places where no risk was identified.

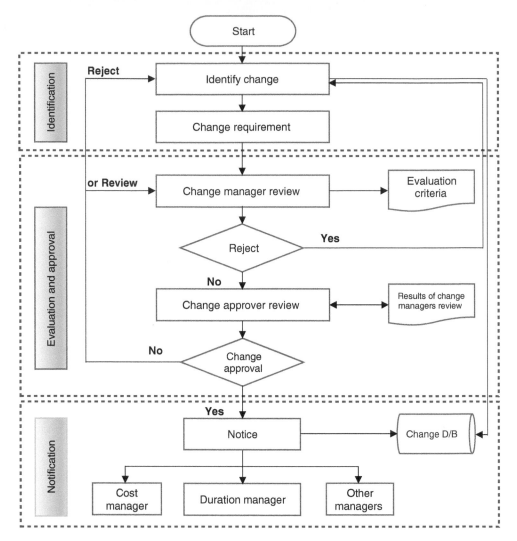

Change management process. Source: Yong Woon Cha et al., Change and contract management modules of intelligent-rogram management information systems (i PgMIS) for urban renewal projects, 2012 Proceedings of the 29th ISARC, Eindhoven, Netherlands.

Change control ensures that change requests are considered and that variations are managed in a controlled way.

Evaluation criteria for changes will include as a minimum, the impact on:

- Scope
- Quality
- Time
- Resource availability
- Whole-life costs
- Benefits
- Risk profile
- Stakeholder perceptions and expectations

Change freeze

Some organisations openly adopt a change freeze, or no change mentality, after a particular phase in the project life cycle, typically after Design. This is done both to discourage changes, which will have a disruptive effect on the project, and also to emphasise the need to get an agreed set of requirements 'up front' such that the complete project scope can be defined. The only time that changes are allowed are when the deliverables would be unsafe, not work or in contradiction to the law or regulation. Having such a rule can drive better planning.

Guidance Note 43   Stakeholder engagement and communication

Guidance Note 6 addressed work necessary to analyse and map stakeholders.

This guidance note addresses different ways of engaging and communicating with stakeholders over the life cycle.

Stakeholder analysis provides initial insight into the power, interests, attitudes and connections between stakeholders. Often such initial insight is based on assumptions rather than reliable knowledge, so the first task of stakeholder engagement (beyond initial analysis and mapping) is to check out assumptions and build a more reliable set of data about the needs, preferences and likely behaviours of stakeholders. A primary responsibility for the client project manager is to build and maintain relationships with stakeholders, so the project is planned and delivered based on sound information.

Stakeholder engagement requires activities being led by the right people, at the right time.

By creating a relationship management plan (RMP) for key stakeholders, you can capture salient points about the stakeholder to help you work and interact with them. The RMP is a summary of the stakeholder, their interest, sentiment toward the project in general. It could also include how they like to be contacted/worked with – their preferred method of contact (calls, emails, face-to-face), frequency of contact and other useful information.

It also helps every team member who may be working with them to have this information at their fingertips. It is also invaluable when there is a change of project personnel and a new team member needs to come up to speed.

The RMPs should, as a minimum, be reviewed every time the stakeholder map is reviewed.

Engagement and communication need to be two-ways – 'transmit and receive' – taking into account the optimal medium for the communication.

Media richness is a concept that describes the ability of the communication method to convey the information accurately. Methods that combine verbal and non-verbal communication are more media rich than methods that rely only on the written word.

The figure shows that a good choice of communication media takes into account the degree to which the situation is clear or ambiguous, for example if the communication is to tell a large number of people about some

Code of Practice for Project Management for the Built Environment, Sixth Edition. Chartered Institute of Building.
© 2022 John Wiley & Sons Ltd. Published 2022 by John Wiley & Sons Ltd.

Guidance Note **43**

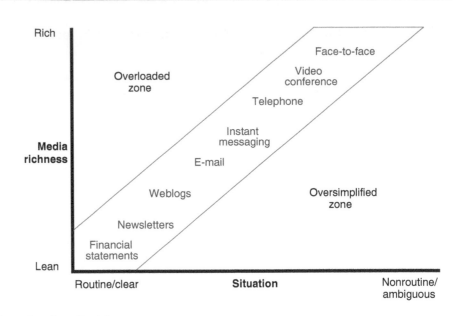

Hierarchy of media richness.

factual data, a 'media rich' method such as a video conference would be inappropriate, whereas if the situation is complex and disputed, a newsletter would not do.

Another way of considering the right choice of communication method is to think about what the stakeholder is intended to do with the information. If the stakeholder needs to be influenced to own an aspect of the project, for example to negotiate additional funding, the investment in time to engage, communicate and influence will be less than if the objective is to ask the stakeholder to endorse an approach.

Direct engagement – the methods chosen to communicate should be set out in a stakeholder and communications plan, which establishes how and when you will communicate with stakeholders, by what methods and who owns the relationships with key stakeholder groups. Outside of traditional media routes of engagement (Social, newsletters etc. . .), you should consider what direct engagement you will have with local communities and key stakeholders.

Options could include Strategic Stakeholder Reference Groups (SRG) – regular forums held with key local authority/environment figures to deliver information and accept feedback.

More locally, Community Liaison Groups (CLG) can be held – which are regular meetings held with local representatives who will feed information back to their

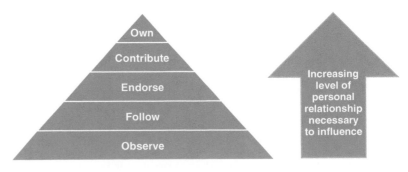

Levels of personal relationship necessary to increase influence.

local communities and are often made up of local business owners, Parish Council members etc.

Where the project requires people to adopt new ways of working or living, it is normal for people to have multiple questions and maybe concerns about the future. It is unhelpful to see this as 'resistance to change' that is in some way illogical or illegitimate. The client project manager must find a way of engaging these stakeholders, listening to their needs and any concerns, and keeping open communication through the life of the project.

Examples may be a funding body, a planning authority, prospective tenants, staff members at a retail outlet or a workers' representative for an SME.

Early engagement can be key. Stakeholders' contributions can enhance the end product and their early involvement can help smooth the path to delivery.

Many stakeholders go on to become customers, if they're not already. Ultimately, stakeholders can make or break a project, but by undertaking early engagement you can work with the concerns and issues of local communities by undertaking that two-way engagement to understand their needs but also to create an understanding about what is within the scope of a project and what is outside of it.

Advocates are a class of stakeholder who can influence many others for you. Social media influencers are probably the best-known type of advocate the general public is aware of in today's society. In the Built Environment, consider for example the ability of lobby groups who may promote or oppose a development, or the power of end users to influence solutions to motorway design.

Identifying advocates, building mutually beneficial relationships and helping them get your messages across can be a distinct activity in itself.

Guidance Note 44 Overview of the UK Health and Safety at Work etc. Act, 1974 (HSWA 1974)

The Health and Safety at Work etc. Act 1974 (HSWA) is the primary piece of legislation covering occupational health and safety in Great Britain.

Section 2(1) of the HSWA sets out the general duty an **employer has to employees** and states: 'It shall be the duty of every employer to ensure, so far as is reasonably practicable, the health, safety and welfare at work of all his employees'.

Section 2(2) extends this, so far as is reasonably practicable, to include the:

- Provision and maintenance of safe plant and safe systems of work

- Arrangements for ensuring safe means of handling, use, storage and transport of articles and substances

- Provision of information, instruction, training and supervision

- Provision of a safe place of work and provision and maintenance of safe access and egress to that workplace

- Provision and maintenance of a safe working environment and adequate welfare facilities

HSWA Section 3(1) is a similar duty for **employers, but in respect of non-employees**. A non-employee can be anyone from workers of another company to members of the public who might be affected by the work activities of the employer, and states: *It shall be the duty of every employer to conduct his undertaking in such a way as to ensure, so far as is reasonably practicable, that persons not in his employment who may be affected thereby are not thereby exposed to risks to their health or safety.*

Section 3(2) imposes a similar duty on the **self-employed** in respect of themselves and others.

Directors and senior managers have a secondary liability under HSWA Section 37 in certain circumstances for the offence committed by an employer (which has the primarily liability). The prosecution must first prove the employer committed a health and safety offence and, secondly, that this breach was due to the Director/Senior Manager's consent or connivance or was attributable to the Director/Senior Manager's neglect.

Code of Practice for Project Management for the Built Environment, Sixth Edition. Chartered Institute of Building.
© 2022 John Wiley & Sons Ltd. Published 2022 by John Wiley & Sons Ltd.

Section 7 of the HSWA can be used to prosecute a wide range of **employees** from frontline workers to senior managers who are not senior enough to be covered by Section 37 of the HSWA. Section 7 states: '*It shall be the duty of every employee while at work to take reasonable care for the health and safety of himself and of other persons who may be affected by his acts or omissions at work . . .*'

Key points to bear in mind in relation to health and safety legislation

There are several important points to note that apply to HSWA and other regulations (such as CDM 2015) made under the HSWA:

- Contraventions of health and safety law are concerned with failures to manage risks to health and safety – they do not require proof the offence caused any actual harm, the contravention is in creating a risk of harm and/or a material breach of law.

- Health and safety responsibilities cannot be delegated or contracted out – the day-to-day activities can be delegated, but the statutory liability remains with the duty holder.

- Employers cannot escape conviction by blaming the exposure to risk on the failures of its employees – employers have duties to inform, train, manage, supervise and monitor their employees.

- Much of the health and safety legislation is underpinned by the term 'reasonably practicable' – this means employers do not have an absolute duty to do something, but can make a judgement.

Reasonably practicability is about weighing the risk against the sacrifice needed to further reduce that risk. The Health and Safety Executive notes the decision is weighted in favour of health and safety because the presumption is the duty holder should implement the risk reduction measure. To avoid having to make this sacrifice, the duty holder must be able to demonstrate it would be grossly disproportionate to the benefits of risk reduction that would be achieved. Thus, the process is not just one of balancing the costs and benefits of measures but, rather, of adopting appropriate control measures except where they are ruled out because they involve grossly disproportionate sacrifices.

Guidance Note 45   Overview of the UK Construction (Design and Management) Regulations 2015 (CDM 2015)

These regulations first came into force in 1995 and the latest version, CDM 2015, came into force on 6 April 2015, supplemented by the L153 Guidance.[1] CDM 2015 expands upon the general duties of the Health and Safety at Work Act 1974 with respect to the management of health, safety and welfare when carrying out construction projects. It sets out what those involved in planning, procurement, design and undertaking the construction work need to do.

CDM 2015 creates five duty holder roles: Client, Principal Designer (PD), Designer, Principal Contractor (PC) and Contractor. Where there is only one Contractor on site, CDM 2015 only requires three duty holder roles to be undertaken: Client, Contractor and Designer. Where there is more than one Contractor on site, CDM 2015 requires all five duty holder roles to be undertaken and, as such, the Client must appoint in writing a PD and PC.

- **Clients** are organisations or individuals for whom a construction project is carried out. They must make suitable arrangements for managing a project including preparing and providing relevant health and safety information and making sure the PD and PC carry out their duties.

- **Principal Designers** are the Designer with control over the pre-construction (design) phase. Their role is to plan, manage, monitor and coordinate health and safety matters during this phase (which continues throughout the construction phase, as there will typically be temporary works to design). They must help the Client in preparing the *Pre-Construction Information* and must prepare, appropriately review, update and revise from time to time to take account of the work and any changes that have occurred and provide the *Health and Safety File*.

- **Designers,** when preparing or modifying designs, must so far as is reasonably practicable to eliminate, reduce or control foreseeable risks that may arise during the construction, maintenance and use of a building (once it is built). As the definition of 'design' in CDM 2015 is so broad, organisations and individuals who make decisions that affect the design (for instance by

---

[1]   https://www.hse.gov.uk/pUbns/priced/l153.pdf

Code of Practice for Project Management for the Built Environment, Sixth Edition. Chartered Institute of Building.
© 2022 John Wiley & Sons Ltd. Published 2022 by John Wiley & Sons Ltd.

modifying the specification to reduce the cost) could attract Designer duties even if they do not consider themselves to be a Designer in the traditional sense.

- **Contractors** are those who do the actual construction work and can be either an individual or a company.

- **Principal Contractors** are the Contractor with control over the construction phase. Their role is to plan, manage, monitor and coordinate health and safety matters during this phase. They must produce a *Construction Phase Plan* before construction starts and review, update and revise it from time to time during construction works. During the project, they must provide the Principal Designer with any relevant information in their possession for inclusion in the health and safety file.

If Project Managers work for any of those duty holder organisations, they will have the statutory CDM 2015 duties associated with that role. If Project Managers do not work for one of the five duty holders, they will not be subject to CDM 2015 duties; however, they will be subject to the duties in the Health and Safety at Work Act and other relevant legislation.

Guidance Note 46 Overview of the UK Dangerous Substances and Explosive Atmospheres Regulations, 2002 (DSEAR, 2002)

DSEAR require employers and the self-employed to control the risks to safety from fire, explosions and substances corrosive to metals to protect people from risks to their safety in the workplace and to members of the public who may be put at risk by work activities.

Dangerous substances are any substances used or present at work that could, if not properly controlled, cause harm to people as a result of a fire or explosion or corrosion of metal. They can be found in nearly all workplaces and include such things as solvents, paints, varnishes, flammable gases, such as liquid petroleum gas (LPG), dusts from machining and sanding operations, dusts from foodstuffs, pressurised gases and substances corrosive to metal.

HSE notes, to comply with DSEAR, employers must:

- Find out what dangerous substances are in their workplace and what the risks are

- Put control measures in place to either remove those risks or, where this is not possible, control them

- Put controls in place to reduce the effects of any incidents involving dangerous substances

- Prepare plans and procedures to deal with accidents, incidents and emergencies involving dangerous substances

- Make sure employees are properly informed about and trained to control or deal with the risks from dangerous substances

- Identify and classify areas of the workplace where explosive atmospheres may occur and avoid ignition sources (from unprotected equipment, for example) in those areas.

Further guidance on is available in the L138 Approved Code of Practice and guidance.[1]

---

[1] Health and Safety Executive (2002) "Dangerous substances and explosive environments code of practice and guidance". Available at https://www.hse.gov.uk/pubns/priced/l138.pdf (accessed 21 May 2021).

Guidance Note 47 Overview of the UK Housing Acts

Historically, the protection of the private sector residential tenancies began with the inception of the Rent Act 1915, which controlled rent in order to meet war-time shortages. Prior to the Rent Act 1915, the bargaining power between landlords and tenants was very uneven and heavily weighted in the landlord's favour.

The Rent Act 1977[1] set up a system for registering 'fair rents' for dwelling-houses. Fair rents are assessed by a rent officer, who is an official with the Valuation Office Agency.

Once a fair rent has been registered, it is the maximum rent a landlord may charge for that property.

Note: if a fair rent is registered after the tenancy was granted, the fair rent will be the maximum payable, even if the tenancy agreement provided for a higher rent.

Housing Act 1985[2]

A major reform of Housing policy was introduced by Margaret Thatcher's government, which allowed certain qualifying tenants the right to buy their homes from the relevant local authority at a discount. In short, it created succession of Council Houses, and it facilitated the transfer of council housing to not-for-profit housing associations.

Historically, the Housing Act 1985 did not apply to Housing Associations; however, in 2015 the government passed legislation, which means Housing Associations are no longer excepted and some Housing Association tenants (who meet certain criteria) now benefit from the right to buy (which is different to the Right to Acquire).

This has led to a shortfall in council and social housing in subsequent years, which has created their own social and economic issues.

The Right to Acquire was a right granted to certain qualifying housing association tenants. The discounts available for a right to acquire as opposed to a right to buy are not as attractive.

[1] UK Government Rent Act 1977 https://www.legislation.gov.uk/ukpga/1977/42/contents
[2] UK Government Housing Act 1985 https://www.legislation.gov.uk/ukpga/1985/68/contents

Code of Practice for Project Management for the Built Environment, Sixth Edition. Chartered Institute of Building.
© 2022 John Wiley & Sons Ltd. Published 2022 by John Wiley & Sons Ltd.

Guidance Note 47

Housing Act 1988[3]

The Housing Act attempted to remove rent control from the private sector and from the 15 January 1989 (the date the 1988 Act came into force), new residential lettings would no longer be subject to their control. The Assured tenancy was a form of letting which permitted the landlord to charge a commercial rent but gave the tenant some security in their home.

Housing Act 1996[4]

The Housing Act 1996 gave greater autonomy to private sector landlords by stipulating that unless the parties agreed otherwise, any assured tenancy granted on or after 28 February 1997 would be an assured shorthold tenancy. This form of tenancy enables a landlord to recover possession on giving notice and denied the tenant any significant security.

Housing Act 2004[5]

The Housing Act 2004 introduced Home Information Packs (which have since been abandoned). It also extended the regulation of houses in multiple occupation. This means any house in multiple occupation needs to be specifically licenced by the local authority.

The 2004 Act also provided a legal framework for tenancy deposit schemes, which are intended to ensure good practice regarding deposits in assured shorthold tenancies and make dispute resolution relating to them easier.

What tenancy and what rights?!

For Project Managers, it is important to know what tenancy and therefore what rights tenants who occupy dwellings have. This will have an impact on timings and also potential financial implications if, for example, a tenant has the benefit of the right to buy. A short table summarising what tenancy applies in what situation is below:

---

[3] UK Government Housing Act 1988 https://www.legislation.gov.uk/ukpga/1988/50/contents
[4] UK Government Housing Act 1996 https://www.legislation.gov.uk/ukpga/1996/52/contents
[5] UK Government Housing Act 2004 https://www.legislation.gov.uk/ukpga/2004/34/contents

Housing Act: types of tenancy

Type of tenancy	Who is it applicable to	Exclusions to the tenancy
Regulated tenancies, i.e. Rent Act tenancies (Rent Act 1977)	• During the contractual tenancy, the tenant is a protected tenant (Section 1 of the Rent Act 1977[1]). • Once the contractual term ends, the tenant becomes a statutory tenant, as long as certain conditions set out in the Rent Act 1977 are met (Section 2 of the Rent Act 1977[2]). NB: The main condition is that the tenant occupies the dwelling-house as their residence (*section 2(1)(a), RA 1977*[3]).	• Certain shared ownership leases • Lettings to students • Holiday lettings • Lettings of licensed premises • Lettings with resident landlords
Assured Tenancies (Housing Act 1988)	A dwelling-house let as a separate dwelling was automatically an assured tenancy if all of the following conditions were met: • The tenant, or each of the joint tenants, was an individual • The tenant, or at least one of the joint tenants, occupied the dwelling-house as their only or principal home • The tenancy was not one that could not be an assured tenancy (see below)	Tenancies that cannot be assured tenancies (or, by definition, Assured Shorthold Tenancies) are: • Tenancies made before 15 January 1989, or pursuant to a contract made before that date (paragraph 1, *Schedule 1, HA 1988*[4]) • Some qualifying high-value properties • Tenancies at a low rent (please refer to the Act) • Business tenancies to which Part II of the Landlord and Tenant Act 1954 applies • Tenancies of premises that are licensed for the supply of alcohol for consumption on the premises • Tenancies where agricultural land of more than two acres is let with the dwelling-house • Tenancies of agricultural holdings or farm business tenancies • Lettings to students by certain landlords, mainly educational institutions • Holiday lettings • Lettings by resident landlords • Crown tenancies • Lettings by specified public bodies, including local authorities
Assured shorthold tenancies	Most tenancies will be assured shorthold tenancies.[5]	
Common law tenancies	A common law tenancy is one that falls outside the statutory regimes and is instead governed by the common law For example, a residential tenancy in England where the yearly rent exceeds £100,000 will also normally be a common law tenancy	Is a tenancy that is neither an Assured Shorthold Tenancy or an Assured Tenancy

(Continued)

(Continued)

[1] UK Government Section 1 of the Rent Act 1977 https://www.legislation.gov.uk/ukpga/1977/42/section/1 (accessed June 2021)

[2] UK Government, Section 2 of the Rent Act 1977 https://www.legislation.gov.uk/ukpga/1977/42/section/2 (accessed June 2021)

[3] UK Government, Section 2 of the Rent Act 1977 https://www.legislation.gov.uk/ukpga/1977/42/section/2 (accessed June 2021)

[4] UK Government, Schedule 1 of Housing Act 1988 https://www.legislation.gov.uk/ukpga/1988/50/schedule/1 (accessed June 2021)

[5] It is advisable to seek professional legal advice in order to determine what type of tenancy the property is subject to in order to prevent any issues arising.

Guidance Note 48 Overview of the UK Town and Country Planning Act 1990

Background

The Town and Country Planning Act 1947 (TCPA 1947) was the first comprehensive piece of legislation to control the development and use of land in England and Wales. Since 1947, the Planning Act has continued (and continues) to evolve to deal with the demands of modern day life.

The government has recently published a White Paper entitled 'Planning for the Future'[1] which will overhaul the planning regime if it is passed by Parliament.

Current legislation

The current incarnation of the Town and Country Planning Act is known as 'Town and Country Planning Act 1990' (TCPA 1990) and came into force on 24 May 1990.

The TCPA 1990 is split into several parts or sections, which will dictate practical information such as what is required in planning applications or at what point a planning permission is implemented.

The TCPA 1990 defines a development *as carrying out of building, engineering, mining or other operations in, on, over or under land, or the making of any material change in the use of any buildings or other land.*[2]

An important consideration for any project manager dealing with the development or material development of property is the date planning permission is implemented. The reason for this is because planning permission will have a 'shelf life' of two or three years (this is dictated by the planning authority), and therefore, once any material operational works* have begun, the planning permission would have been implemented and time will start running.

In addition, reference data can be obtained from The Town and Country Planning (Development Management Procedure and Section 62A Applications) (England) (Amendment) Order 2021, to ensure that fire safety matters, as they relate to land use planning, are incorporated at the planning stage for schemes involving a high-rise residential building, i.e. a building above 18 meters in height.

---

[1] UK Government "Planning for the Future: planning policy changes in England in 2020 and future reforms (2021)". https://commonslibrary.parliament.uk/research-briefings/cbp-8981/ (accessed June 2021).

[2] UK Government, Section 55 of the Town and Country Planning Act 1990 https://www.legislation.gov.uk/ukpga/1990/8/section/55?view=plain (accessed June 2021).

Code of Practice for Project Management for the Built Environment, Sixth Edition. Chartered Institute of Building.
© 2022 John Wiley & Sons Ltd. Published 2022 by John Wiley & Sons Ltd.

**Material operational works* include:

(a) any work of construction in the course of the erection of a building;

    (aa) any work of demolition of a building;

(b) the digging of a trench which is to contain the foundations, or part of the foundations, of a building;

(c) the laying of any underground main or pipe to the foundations, or part of the foundations, of a building or to any such trench as is mentioned in paragraph (b);

(d) any operation in the course of laying out or constructing a road or part of a road;

(e) any change in the use of any land which constitutes material development.

Who can apply for planning permission?

Anyone can apply for planning permission on any parcel of land. The planning application (and subsequent planning permission) *attaches to the land* not to the person making the application. Project managers will need to consult with various planning consultants at all aspects of the planning cycle including any pre-application matters, which could include the design and feasibility studies.

Failure to comply

Failure to comply with planning conditions stated in the planning permission itself or with the TCPA 1990 can lead to:

1. Enforcement action by the appropriate authority;

2. Injunctions;

3. Fines/compensation/imprisonment.

It is therefore vital for project managers to ensure they have complied with and sought confirmation for the relevant planning authority that the planning conditions contained within a planning permission have been discharged or complied with by receiving written confirmation. This will be used as evidence should there be any dispute in the future.

Planning obligations under Section 106 of the TCPA 1990[3]

Section 106 Agreements and unilateral lateral undertakings are often colloquially called Planning Obligations. The purpose of these planning obligations is to enable the relevant planning authority to maintain some form of control of the type of development, as well as acting as a direct contract with the landowner (and developer, if applicable). They are also a means of re-claiming funds to put towards the additional costs that will be incurred once the development has been constructed for education, transport and affordable housing amongst

---

[3] UK Government, Housing Grants, Construction and Regeneration Act 1996, Section 106 https://www.legislation.gov.uk/ukpga/1996/53/part/II/crossheading/introductory-provisions (accessed June 2021).

other matters. It is important to factor these costs into feasibility studies in the pre-application stage of any planning permission.

It is important to note that planning obligations *bind the land*.

Latest regulations for general permitted development due to Covid-19

The General Permitted Development Rights are deemed rights granting planning permission for certain developments without the need to make a formal application for planning permission.

Due to the impact of Covid-19 in 2020, the Town and Country Planning (General Permitted Development) (England) (Amendment) 2020 was published on the 9 November 2020 ('2020 Regulations') and came into force on the 1 April 2021. The 2020 Regulations make a number of changes to the Town and Country Planning (General Permitted Development) (England) Order 2015. For example, a number of temporary permitted development rights such as the right for restaurants, cafes etc. to provide takeaway food until 23 March 2022. Note: After 23 March 2022, the premises will have to revert back to its original use.

For details of the current Act, please see: www.legislation.gov.uk.

Guidance Note 49   Implications of the Housing Grants, Construction and Regeneration Act 1996, Amended 2011

The *Housing Grants, Construction and Regeneration Act 1996* (known as the 'Construction Act') (as amended by the *Local Democracy, Economic Development and Construction Act 2009*) has been an important piece of legislation affecting construction contracts since it came into force on 1 May 1998.

Does the Construction Act apply?

The first important point to consider when dealing with the Construction Act is whether or not it actually applies to a particular contract. The Construction Act will automatically apply to construction contracts when formed and applies equally to written and oral contracts. For the purposes of this legislation, the term 'construction contract' means an agreement for any of the following:

- carrying out 'construction operations';

- arranging the carrying out of 'construction operations' by others; or

- providing labour, or the labour of others for the carrying out of 'construction operations'.

The definition includes agreements for architectural or surveying works, and advice on building including interior or exterior decoration and landscaping where they relate to 'construction operations'.

The meaning of 'construction operations' has a broad scope and covers many aspects of construction works. There are also several types of contract expressly excluded from the scope of the legislation such as, for example, drilling for oil or the extraction of natural gas. The full definitions of 'construction contracts' and the exclusions are included in sections 104 and 105 of the Construction Act 1996.[1]

It is also important to note, a contract is not a construction contract (and therefore the Construction Act does not apply) if:

- it relates to works on a person's home in which they live or intend to live;

- it relates to works outside of England, Wales or Scotland;

- the contract is an employment contract.

---

[1] UK Government, Housing Grants, Construction and Regeneration Act 1996, Sections 104 and 105 https://www.legislation.gov.uk/ukpga/1996/53/part/II/crossheading/introductory-provisions (accessed June 2021).

Code of Practice for Project Management for the Built Environment, Sixth Edition. Chartered Institute of Building.
© 2022 John Wiley & Sons Ltd. Published 2022 by John Wiley & Sons Ltd.

Project managers of these construction contracts need to be familiar with the statutory payment mechanisms included in the act to ensure their compliance. Failure to comply with certain aspects of the Construction Act 1996 can result in disputes arising between the parties, defaults on payments and substantial delays to works.

Key concepts – payment terms

The Construction Act sets out several key contractual payment terms a construction contract must have in it and certain requirements in respect of those payment terms. Where a construction contract does not comply, terms are then implied (by virtue of the Scheme for Construction Contracts 1998[2]) into the contract to replace non-compliant or missing terms.

It is important for a project manager to understand the payment provisions in any construction contract in respect of which they are appointed, as well as understanding the legal requirements under the Construction Act, so they are aware of any payment terms that may be implied into or override the terms of the contract itself.

In addition, various notices are required to be provided at certain times in respect of payments under construction contracts and the project manager may ultimately be responsible for issuing or reviewing any such notices. It is therefore important they understand when notices are required and what they need to contain.

The key requirements in terms of payment are:

- A right to payment to the contractor in instalments, staged payments or other periodic payments where the works are expected or specified to last more than 45 days.

- Having an adequate payment mechanism specifying what payments become due, when they become due, and giving a final date for payment (the parties are free to agree the length between the due date and the final date for payment).

- Not having a clause specifying that payment will be made when received from a third party, or a 'pay-when-paid' clause (subject to certain exceptions where insolvencies occur) nor can it be conditional on (subject to certain limited exceptions) performance of obligations under another contract or a decision by another person in respect of another contract.

- The provision of a payment notice within five days of the payment due date. This can be given by the payer, the payee or some other specified person. Importantly, the payment notice must set out the amount due as at the due date *and* the basis on which that amount is calculated. There is also a mechanism relating to default payment notices where a notice is not provided in accordance with this requirement (see section 110B of the Construction Act[3]).

[2] UK Government, Scheme for Construction Contracts, 1998 https://www.legislation.gov.uk/uksi/2011/2333/contents/made (accessed June 2021).

[3] UK Government, Housing Grants, Construction and Regeneration Act 1996, section 110 https://www.legislation.gov.uk/ukpga/1996/53/part/II/crossheading/payment (accessed June 2021).

The references to due date and final date for payment can be confusing. Essentially the due date is when the obligation to pay crystallises and the amount due is determined as at that date. The final date for payment is the date on which the payment is actually required to be made.

The payer is required to pay the amount in a payment notice on or before the final date for payment even if they dispute the payment or the amount of it, unless a valid pay less notice has been provided in accordance with the requirements of the contract.

It is important the project managers who are required to provide any payment notices comply with the notice requirements of the particular contract, including how the notices are sent to the other party.

Generally, parties are free to agree the amounts, intervals and circumstances of payments becoming due in accordance with these requirements. Where this has not been done, or the requirements have not been included, the Construction Act 1996 specifies that terms from the *Scheme for Construction Contracts* are then implied (note that a different scheme applies in respect of Scotland). Under this scheme:

- Valuations should be conducted based on the value of work performed during the 'relevant period', which is specified as every 28 days (where no such period is specified in the contract).

- Payment becomes due seven days after the relevant period or the making of a claim for payment by the payee (whichever is later) and with a separate time period for the final payment due under the contract.

- The final date for making payment is 17 days from the date the payment becomes due.

- Any pay less notices must be given not later than seven days before the final date for payment.

Key concepts – adjudication

Under the Construction Act 1996, a construction contract should also provide the parties with a right to refer disputes arising for adjudication. Where this is not complied with, a term to the same effect is then implied, meaning a party to a construction contract will always have the option of adjudication and the Scheme for Construction Contracts will imply various terms in respect of such adjudication.

Non-compliance

Where a payment becoming due under the payment mechanism is not paid, or any part remains outstanding, by the final due date, the unpaid party has the right to suspend any or all of its obligations under the contract by providing not less than seven days' notice.

This also then entitles the unpaid party to any additional time involved in suspending and remobilising the works under the contract, and they are entitled to the reasonable costs and expenses of the suspension.

Project managers should therefore ensure compliance with the payment provisions in the Construction Act 1996 to prevent possibility of suspension of works, which could result in a costly delay to construction projects.

For details of the current Act, see Housing Grants, Construction and Regeneration Act 1996[4] noting the amendments as made by Local Democracy, Economic Development and Construction Act 2009.[5]

4   UK Government Housing Grants, Construction and Regeneration Act 1996 https://www.legislation.gov.uk/ukpga/1996/53/contents (accessed June 2021).

5   UK Government Local Democracy, Economic Development and Construction Act 2009 https://www.legislation.gov.uk/ukpga/2009/20/contents (accessed June 2021).

Guidance Note **49**

Guidance Note 50   Overview of the Building Safety Bill (2021) and the UK Fire Safety Act (2021)

Building Safety Bill (2021)

'The objectives of the [Building Safety] Bill, published in July 2021, are to learn the lessons from the Grenfell Tower fire and to remedy the systemic issues identified by Dame Judith Hackitt by strengthening the whole regulatory system for building safety. . .'

. . .ensuring there is greater accountability and responsibility for fire and structural safety issues throughout the lifecycle of buildings in scope of the new regulatory regime for building safety. Building Safety Bill – Explanatory Notes [3] and [4][1]

We are determined to learn the lessons from that fateful night at Grenfell Tower and ensure that a tragedy like this does not happen again – Lord Greenhalgh, Building Safety and Fire Minister

In July 2017, the Government announced an Independent Review of Building Regulations and Fire Safety led by Dame Judith Hackitt. The final report of the Independent Review of Building Regulations and Fire Safety entitled Building a Safer Future was published on 17 May 2018[2] and identified four issues.

- Ignorance
- Indifference
- Lack of clarity on roles and responsibilities
- Inadequate regulatory oversight and enforcement tools

These issues have helped to create a cultural issue across the sector, which can be described as a 'race to the bottom'. . .there is insufficient focus on delivering the best quality building possible, in order to ensure that residents are safe, and feel safe – Dame Judith Hackitt

The report made 53 recommendations and called on Government to:

- Create a more effective regulatory and accountability framework to provide greater oversight of the building industry
- Introduce clearer standards and guidance

[1] UK Government, Building Safety Bill Explanatory Notes https://assets.publishing.service.gov.uk/government/uploads/system/uploads/attachment_data/file/901869/Draft_Building_Safety_Bill_PART_2.pdf
[2] UK Government "Building a Safer Future" 2018 https://assets.publishing.service.gov.uk/government/uploads/system/uploads/attachment_data/file/707785/Building_a_Safer_Future_-_web.pdf

- Put residents at the heart of a new system of building safety

- Help to create a culture change and a more responsible building industry.

Contents of the Building Safety Bill[3]

Part 1 – Introduction

Provides an overview of the Bill, which contains six parts and nine sections, and makes a number of changes to existing legislation, most notably the Building Act 1984.

Part 2 – The Regulator and its functions

Establishes a Building Safety Regulator within the Health and Safety Executive, with functions in relation to buildings in England. It also contains definitions of 'building safety risk' and 'higher risk building'.

Part 3 – Building Act 1984

Amends the Building Act 1984 as it applies to England and Wales. Provides that the building safety regulator (a) is the building control authority in relation to higher risk buildings and (2) must establish and maintain registers of building control approvers and building inspectors to improve competence levels.

Part 4 – Higher risk buildings

It defines the scope of the regime for higher risk buildings in occupation. It defines and places duties on the Accountable Person (the dutyholder in occupation) in relation to building safety risks in their building.

Part 5 – Supplementary and general

(a) provision requiring new homes ombudsman scheme to be established.

(b) powers to make provision about construction products.[4]

(c) provision about the regulation of architects.

Part 6 – Technical clauses related to the Bill

Two draft statutory instruments also published alongside the Bill:

- The Building (Appointment of Persons, Industry Competence and Dutyholders) (England) Regulations [2021]

- The Higher Risk Buildings (Descriptions and Supplementary Provisions) Regulations [2021]

Schedules

Schedule 1 – Amendments of the Health and Safety at Work etc. Act 1974

Schedule 2 – Authorised officers: investigatory powers

Schedule 3 – Co-operation and information sharing

---

3 UK Government, https://www.gov.uk/government/collections/building-safety-bill
4 New national construction products regulator to be established within the Office of Product Safety and Standards.

Schedule 4 – Transfer of approved inspectors' functions to registered building control approvers

Schedule 5 – Minor and consequential amendments in connection with Part 3

Part 1 – Amendments of the Building Act 1984

Part 2 – Other amendments

Schedule 6 – Appeals and other determinations

Schedule 7 – Building Safety Charges

Schedule 8 – The New Homes Ombudsman scheme

Schedule 9 – Construction products regulations

Explanatory notes

These do not form part of the Bill but explain what each part of the Bill will mean in practice; provide background information on the development of policy; and provide additional information on how the Bill will affect existing legislation in this area.

Buildings in scope

The Bill applies to **higher risk buildings**, which are defined as multi-occupancy residential and student accommodation buildings of 18 m and above (over seven storeys) in height.

The Bill gives the Secretary of State the power to amend the definition of a higher risk building in light of Government research or on the basis of evidence and advice from the Building Safety Regulator (*the Bill is enabling legislation, secondary legislation will follow*).

Building safety regulator

A Building Safety Regulator will be established in the Health and Safety Executive (HSE) to provide oversight of the new system and with powers of enforcement and sanctions.

The Regulator will have powers to prosecute all offences in the Bill and the Building Act 1984 and to issue **compliance notices** (requiring issues of non-compliance to be rectified by a set date) and **stop notices** in design and construction (requiring work to be halted until serious non-compliance is addressed).

Failure to comply with compliance and stop notices will be a criminal offence, with a maximum penalty of up to two years in prison, an unlimited fine (*including a breach of registration requirements*) or both. This also applies to the requirement to register a building and apply for and display a Building Assessment Certificate.

Dutyholders

During Design and Construction: the SI[5] published alongside the Bill sets out the framework of duties for those persons and organisations ('dutyholders') who commission the building work (the Client), and undertake the design

---

[5]   The Building (Appointment of Persons, Industry Competence and Dutyholders) (England) Regulations [2021]

(**Principal Designers**, **designers**) and construction/refurbishment work (**Principal Contractor**, **contractors**). This includes duties of cooperation, coordination, communication and competence.

During occupation: the **Accountable Person** is defined as (a) a person who holds a legal estate in possession in any part of the common parts or (b) a person who is under a relevant repairing obligation in relation to any part of the common parts. The Accountable Person is responsible for:

- registering the building before it becomes occupied and applying for a **Building Assessment Certificate** within 6 months of occupation,

- appointing a person as the **Building Safety Manager**,

- assessing the building safety risks relating to the building and demonstrating how they are meeting this ongoing duty via their safety case and Safety Case Report[6],

- preparing a residents' engagement strategy for promoting the participation of residents in the making of building safety decisions.

The Building Safety Manager is responsible for managing the building in accordance with the **Safety Case Report**.

The Bill also introduces the role of **Principal Accountable Person** where there is more than one accountable Person.

Golden thread

The Bill includes provisions to require the creation and maintenance of a golden thread of information. The intention is to ensure the right people have the right information at the right time to ensure buildings are safe and building safety risks are managed throughout the building's life cycle.

- In May 2021, the '**definition and principles**' drafted by the Building Regulations Advisory Committee's (BRAC) Golden Thread Working Group received ministerial approval.

- The **definition** makes clear that the golden thread covers the information, documents and information management processes used to support building safety.

- **Information** is defined as 'all the information necessary to understand and manage risks to prevent or reduce the severity of the consequences of fire spread, or structural collapse in a building'.

- As a '**single point of truth**', the golden thread 'will record changes, including the reason for change, evaluation of change, date of change and the decision-making process' and who made the changes.

Gateways

There is a requirement to demonstrate building safety though a new system of gateway points during design and construction, and a Safety Case Report during occupation.

---

[6] Safety case principles for high rise residential buildings: Building safety reform Early key messages (https://www.hse.gov.uk/building-safety/news/safety-case-principles.pdf)

- Gateway One (Planning) – ensures that fire safety matters, as they relate to land use planning, are incorporated at the planning stage for schemes involving a high-rise residential building.

- Gateway Two (prior to construction work beginning) – will occur prior to construction work beginning on a higher risk building and replaces the current building control 'deposit of plans' stage. It provides a 'hard stop' where construction cannot begin until the Building Safety Regulator is satisfied that the design meets the functional requirements of the building regulations.

- Gateway Three (at the final certificate/completion stage) – is equivalent to the current completion/final certificate phase, where building work on a higher risk building has finished and the Building Safety Regulator assesses whether the work has been carried out in accordance with the building regulations. Only once Gateway three has been passed can the new building be registered with the Building Safety Regulator and occupation of the building allowed to commence.

At Gateway three, the Client, Principal Designer and Principal Contractor will also be required to produce and co-sign a final declaration confirming that to the best of their knowledge the building complies with building regulations. [Impact Assessment]

Competency

The Building Safety Regulator will have a duty to establish a new **industry competence committee** to advise on industry competence, oversee the longer-term development of the competence frameworks and drive improvements in levels of competence.

The Bill creates powers to prescribe the competence requirements on the Principal Designer and Principal Contractor and to impose duties on the persons appointing them to ensure they meet the competence requirements.

When the client choses the 'design and build' procurement route. . .the contractor will commonly (although not always) be appointed both Principal Designer and Principal Contractor. The Client should be satisfied that the contractor has sufficient skills, knowledge, experience and behaviours (competence) to undertake both.

Building control

The Building Safety Regulator will become the building control authority for higher risk buildings, with power to oversee and report on the performance of building control bodies [amends the Building Act 1984].

It also sees the separation of those who 'inspect' from those who 'approve', which was one of the recommendations in the 'Building a Safer Future' report.

Mandatory occurrence reporting

Provisions are included in the Bill that require mandatory occurrence reporting (MOR) to be undertaken for higher risk buildings. All structural and fire safety occurrences that could cause a significant risk to life safety will need to be reported to the Building Safety Regulator.

Dutyholders in design and construction will be required to establish a framework and process for reporting mandatory occurrences, which must include enabling workers on site to report mandatory occurrences.

In occupation, the Accountable Person will be required to set up a framework and process to capture and report mandatory occurrences to the Building Safety Regulator.

Residents

It places a requirement on the Accountable Person to prepare a **residents' engagement strategy** to promote the participation of residents in the making of building safety decisions.

It also places a duty on residents not to act in a way that creates a significant risk of a building safety risk materialising, not to interfere with the relevant safety item.

> *. . . the Bill also includes new provisions that (a) increase the period that residents can make a claim under the Defective Premises Act 1972 by extending the limitation period from 6 to 15 years and (b) require developers to join and remain members of the New Homes Ombudsman scheme, which will require them to provide redress to a homebuyer.*

UK Fire Safety Act (2021)

The Fire Safety Act 2021 received Royal Assent on 29 April 2021 and applies to England and Wales.

It amends the Regulatory Reform (Fire Safety) Order 2005 (FSO) to implement the recommendations of Phase 1 of the Grenfell Tower Inquiry and strengthen the regulatory framework for how building control bodies consult with Fire and Rescue Authorities.

It clarifies that the FSO applies to the structure, external walls (including cladding and balconies) and individual flat entrance doors in buildings which contain two or more domestic premises. This puts the Advice Notes published by Government following the Grenfell Tower fire into legislation.

Owners and managers of multi-occupied residential buildings need to ensure that the fire risk assessment for such buildings are reviewed and updated to encompass the structure, external walls and flat entrance doors.

It imposes new duties on the Responsible Person (RP) to:

- be defined as the person having control of the building,

- be identified, to provide local Fire and Rescue Services with information about the design of the building's external walls and details of the materials they are constructed from,

- to provide specific fire safety information to residents (similarities with the Building Safety Bill.

Guidance Note 51   Design management fundamentals

Good design is fundamental to sustainable development. It plays a crucial role in how places are perceived, with the quality of the built environment contributing to a positive perception. Decisions made about design can affect safety, the environment, ecology, the customer journey, congestion, the economy and wider society.

Design Management identifies people as the real value generators. A design management mindset reinforces good project management, enhancing collaboration and synergy between the design and the business outcomes to assure effectiveness, efficiency and economy (value for money) is achieved from the design and consequently the project.

Purpose

Issues in design can become problematic or cause problems down the line in construction if not proactively identified. The later they are identified, the more time and money is needed to resolve them. This guidance note addresses how to identify and overcome those problems and better still, avoid them by focusing on the fundamentals of design management.

1. Design must be done carried out in accordance and complying with the Construction Design and Management (CDM) Regulations. An overview of the regulation is provided in Guidance Note 45. Also the design for a building must comply with the Building Regulations and other related legislation, approved documents and guidance being drafted under the draft Building Safety Bill as part of the Building Act 1984.

2. Design management is not just about managing the design process, it overtly connects design and business outcomes. It is about empowering the team to ask the right questions and doing the right things at the right time to get the right outcome.

3. Design reviews provide the opportunity to challenge the design and ensure that it will meet needs and benefits for an acceptable whole-life cost. For example, understanding why one design solution might be better compared with another is the responsibility of the designer; however, the client project manager must always represent the operator and end-user viewpoint.

4. If an opportunity for innovation arises, discuss the art of the possible as early as you can and explore whether the design solution has been applied

Code of Practice for Project Management for the Built Environment, Sixth Edition. Chartered Institute of Building.
© 2022 John Wiley & Sons Ltd. Published 2022 by John Wiley & Sons Ltd.

elsewhere so learning on matters related to safety, buildability, quality, cost and time outcomes for your project can be evaluated.

5. Design once, use many times. Don't let designers reinvent elemental design solutions for a familiar problem, for example, design of doorways and doors in an hospital or school.

6. Ensure design information is collated and recorded in accordance with the requirements of the Employer's Information Requirements (EIR). Set up early handover meetings to discuss and agree handover documentation with team members. Consider sending 'sample' handover deliverables at an early stage to establish an acceptable standard by the design authority.

Guidance Note 52   Detailed design

The client project manager is responsible for bringing together the design team, often under the leadership of a design manager or design coordinator, including all consultants hired to bring specific advice, to bring them up to date with all aspects of the project to date and to ensure that design output requirements are understood in line with overall information management requirements (see Guidance Note 10), i.e. the detailed design models and drawings that are required.

The design stage of the project life cycle is usefully seen as a project in itself and in many projects will take many years, involving a complex supply chain. The project manager for this activity has several responsibilities to coordinate the design team activities shown in the figure below, including:

- Monitoring progress, resources and productivity against the design management plan in association with the team.

- Incorporating into the project timeplan the dates for the submission of design reports and periods for their consideration and approval.

- Commissioning specialist reports, for example relating to the site, legal opinions on easements and restrictions and similar matters.

- Ensuring compliance with all relevant regulations, including but not limited to CDM (see also Guidance Note 45).

- Arranging for the team to be provided with all the information it requires from the client in order to execute its duties.

- Coordinating the activities of the various (and sometimes numerous) participants in the total process, for example solicitors, accountants, tax consultants, development consultants, insurance brokers and others may all be involved in the design stage.

- Obtaining regular financial/cost reports and monitoring against budget/cost plans. Initiating remedial action within the agreed brief if the cost reports show that the budget is likely to be exceeded.

- Preparing 'schedule of consents' with action dates, submission documents, status etc., and monitor progress.

- Checking that professional indemnity insurance policies are in place and remain renewed on terms that accord with conditions of engagement.

Code of Practice for Project Management for the Built Environment, Sixth Edition. Chartered Institute of Building.
© 2022 John Wiley & Sons Ltd. Published 2022 by John Wiley & Sons Ltd.

Guidance Note **52**

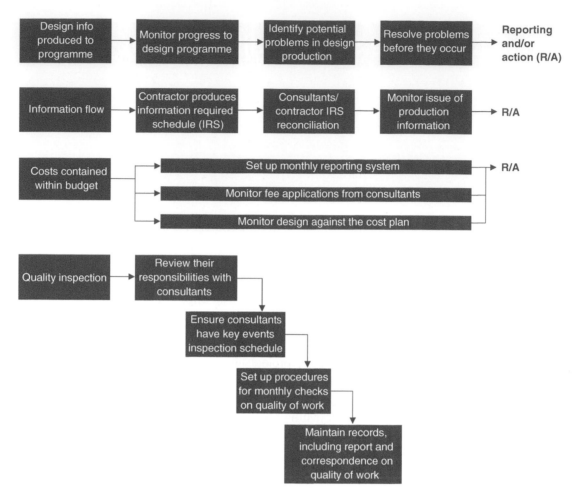

Design team activities.

- Ensuring the 'audit trail' of information is in place in line with the information requirements specified in contracts.

The above flow chart clarifies a standard model of design team activities that can be adapted dependent on the complexity of the project. An essential element is the final sub-flow section regarding quality inspections, which are essential to validate that the design activities have been carried out correctly.

Historically in traditional construction, the majority of the build was carried out on-site and sequentially. Due to improvements and expectations related to quality control and speed of construction, off-site manufacturing and construction is becoming more prevalent. Another important benefit in off-site construction is that the process is carried out in an environmentally friendly and temperature-controlled environment. There is no limit on the type of projects that can adopt off-site manufacturing, and there is now a substantial off-site manufacturing supply chain in existence that can be approached to offer expert advice and costings related to this process.

It is often necessary to carry out a full benefit value and cost analysis prior to deciding on which procurement process is best for the project. A proven benefit of off-site manufacturing is that performance testing of final products can be carried out effectively and efficiently prior to delivery to site. There are, however, logistical issues that will need to be considered with regards to transportation of off-site components and the installation of these into the existing fabric. The testing regimes are particularly relevant and pertinent to mechanical and electrical components and building management and control systems.

Transportation should be carefully investigated because, for example, permits and licenses may be required for large components (such as facades and large mechanical components).

A current advancement in off-site manufacture includes P-DFMA (Platform design) in which building products and components are prepared in a way that enables them to be produced on a large scale and then put together in one place. The use of a set of digitally designed components across multiple types of built asset that are then used wherever possible can minimise the need to design bespoke components for different types of asset. For example, the same component could be used in the construction of a school, hospital and prison.

There are also benefits in health and safety, where off-site manufacture can be carried out in a much more controlled and safe environment.

Further examples of off-site manufacture could include the following:

- Parts of track and parts of rolling stock could be made off site and latterly 'connected' on site;

- External cladding to station facades, staircases, lifts and escalators could be built off site in their entirety and installed without the need for work on-site, other than installation.

- Toilets, waiting rooms, staff accommodation and ticket offices could be constructed as pods off site.

- Repetitive floor plates, whether low rise or high rise, can lead to the effective use of modern methods, including single items such as cladding, stairs, entrance porches and outbuildings, garages and conservatories.

- Pods are commonly used in university student accommodation to contain bathrooms and the like.

- In refurbishment works, there will always be some elements of new build within the overall scheme, for example the structural shell and core may remain in place with differing elements manufactured offsite and installed on site.

- Small projects could potentially be provided as a pre-manufactured unit as a complete entity, or via a combination of pre-made pods.

It is important in the concepts stage of the project that key decisions are made regarding the client requirements for off-site versus on-site installation. In many cases, off-site manufacturers have their own design resource that converts the concepts into detailed design, which ensures a fully coordinated and designed product. The project manager should therefore produce a schedule of components that all stakeholders agree should be produced off site. This should be a live document, which should be a part of the Project Execution Plan and can be regularly updated to monitor assumptions made and value benefits achieved.

Further information can be found in the UK government publication on Modern Methods of Construction[1] that provides a framework of seven categories of innovative construction techniques. These are applied to the residential housing market but have wider application.

---

[1] UK Government (2019) "Modern Methods of Construction: Introducing the MMC Definition Framework". Available at https://www.gov.uk/government/publications/modern-methods-of-construction-working-group-developing-a-definition-framework (accessed 1 April 2021).

Guidance Note 54 Operations and maintenance (O&M) manual indicative content

The Operations and Maintenance (O&M) manual must provide all relevant information to the Operator of the asset based on the intended use and the client's strategy for life cycle management.

The O&M manual will include, as a minimum, information about:

- design principles
- materials used
- as-built drawings and specifications
- asset registers of plant and equipment
- operations and maintenance plans and instructions
- user guidance
- health and safety information
- testing results
- guarantees, warranties and certificates
- assumptions on whole-life costs
- any requirements for demolition, decommissioning and/or disposal

The O&M manual should be considered to be a companion guide to all operators and users of the built asset. It should prioritise health and safety as a core principle in any operations and maintenance regimes. The language of the document should be clear, concise and uncomplicated so that stakeholders from a construction and non-construction background have clarity on how to operate and maintain the facilities.

More complex projects, for example hospitals or research and development facilities, will require more detailed information than that listed above and ultimately the clients project manager needs to manage all design and construction stakeholders to prepare and issue their relevant documentation to be incorporated into the O&M manual.

It is imperative that commercial and contractual arrangements include milestones for delivery of documentation and also clear specifications relating to content requirements, which should be clarified and agreed with the end-user early in the construction life cycle.

Code of Practice for Project Management for the Built Environment, Sixth Edition. Chartered Institute of Building.
© 2022 John Wiley & Sons Ltd. Published 2022 by John Wiley & Sons Ltd.

Guidance Note 55   Client handover checklist: indicative content

- All works complete as per the contract drawings and approved variations
  - Asset configuration, layout & groundworks
  - All appropriate approval certification
  - All appropriate test certification
  - Site/asset cleared of materials and cleaned to agreed standard
- The following systems to be checked for correct operation/function:
  - Mechanical systems
    - Fire suppression
    - Ventilation
    - Air conditioning
    - Heating (space and water)
    - Gas/air/medical gases
    - Plumbing
    - Drainage (internal & external)
  - Electrical systems
    - Fire
    - Security
    - Access controls
    - Lighting (emergency/general/specialist)
    - Power (including Photovoltaic/HV/LV/UPS/back-up generators)
    - Telecoms
    - Data
    - TV/cable/satellite
    - Public address/music
    - Projectors/AV screens
    - Digital advertising/display boards

Code of Practice for Project Management for the Built Environment, Sixth Edition. Chartered Institute of Building.
© 2022 John Wiley & Sons Ltd. Published 2022 by John Wiley & Sons Ltd.

- Specialist equipment/systems
  - Radio communications
  - Highway/traffic controls
  - Access & safety
  - Medical/clinical/scientific
  - Escalators/lifts
  - Commercial ovens/kilns
  - Commercial fridges/freezers
- All finishes satisfactory both internally and externally
  - Ceilings & integrated fittings
  - Wall surfaces
  - Joinery
  - Floor finishes
  - Fixtures
  - Loose furniture
  - Signage & graphics
- Handover of keys (where appropriate)
- List of defects/outstanding works
- Confirm as-built drawings represent as built
- Confirm O&M Manual received
- Provision of updated Health and Safety file
- Instruct/Issue Practical Completion Certificate
- Confirm routine (day to day) maintenance tasks with end user
- Confirm routine (periodic) servicing agreements with appointed provider(s)
- Confirm dates for defects period/inspection(s)
- Instruct outstanding contract payments (less retentions)
- Receipt of spares (where listed in the contract)
- End user training/instruction
- TM 44 register and F gas register
- Updated asbestos register as appropriate
- Confined space schedule/drawing location
- DSEAR/hazardous areas
- Energy certificate as appropriate
- Facilities that dispense fuel to vehicles (filling station). Fitness for purpose certificate

- Above and below ground ancillary fuel storage facilities including transfer/distribution system supplying generator or boiler installations. Fitness for purpose certificate.

- Gas main distribution pipework and certification

- Biological decontamination certs if applicable

- Room integrity tests

- Fire cause and effect matrix

Guidance Note 56   Client commissioning checklist – building services example

The following Building Services systems would typically be commissioned by the manufacturer, installer and/or an independent specialist on behalf of the client and prior to handover of the project to the operator:

- **Mechanical systems**
  - Fire suppression
    - Cause and effects
    - Correct operation of systems
    - Correct interfacing with adjacent systems
    - Connectivity to direct control centres
  - Ventilation
    - Air flow and pressure balancing
    - Interfacing with fire systems
  - Air conditioning
    - Temperature stability
    - Refrigerant types, pressures & quantities
  - Heating (space and water)
    - Operation of gas or oil burning boilers
  - Gas/air/medical gases
    - Testing of pressure vessels
    - Correct flow of gases
  - Plumbing
    - Operation of macerators and internal pumping systems
    - Water softeners
    - Chemical dosing values
  - Drainage (internal & external)
  - Operation of external pumping stations

Code of Practice for Project Management for the Built Environment, Sixth Edition. Chartered Institute of Building.
© 2022 John Wiley & Sons Ltd. Published 2022 by John Wiley & Sons Ltd.

- **Electrical systems**
  - Fire
    - Cause and effects
    - Correct operation of systems
    - Correct interfacing with adjacent systems
    - Connectivity to direct control centres
  - Security
    - Correct operation of systems
    - Correct interfacing with adjacent systems
    - Connectivity to direct control centres
  - Access controls
    - Correct operation of systems
    - Correct interfacing with adjacent systems (fail safe/fail secure)
    - Connectivity to direct control centres
  - Lighting (emergency/general/specialist)
    - Correct operation of systems (automated lux settings, daylight adjustments, colour level adjustment)
    - Correct interfacing with adjacent systems (IOT, emergency lighting, solar shading)
  - Power (including Photovoltaic/HV/LV/UPS/back-up generators)
    - Fire cause and effects
    - Correct operation of systems
    - Correct interfacing with adjacent systems (i.e. mains to back-up switchover systems)
    - Connectivity to direct control centres
  - Telecoms
    - Correct operation of systems
    - Correct interfacing with adjacent systems (automated call systems: fire, police)
    - Connectivity to direct control centres
  - Data
    - Correct operation of systems
    - Correct interfacing with adjacent systems
    - Cable and port termination testing (i.e. fluke or similar)
  - TV/cable/satellite:
    - Correct operation of systems

- Range of channels/bandwidths
- Signal strength/signal quality

- Public address/music
  - Correct operation of systems
  - Correct interfacing with adjacent systems (pre-recorded messages i.e. fire evacuation)
  - Connectivity to direct control centres

- Projectors/AV screens
  - Correct operation of systems
  - Correct interfacing with adjacent systems
  - Connectivity to direct control centres

- Digital advertising/display boards
  - Correct operation of systems
  - Correct interfacing with adjacent systems (pre-recorded messages, i.e. fire evacuation)
  - Connectivity to direct control centres

- **Specialist equipment/systems**
  - Radio communications
    - Correct operation of systems
    - Range of channels/bandwidths
    - Signal strength/signal quality

  - Highway/traffic controls
    - Correct operation of systems
    - Correct interfacing with adjacent systems
    - Connectivity to direct control centres

  - Access & safety
    - Correct operation of systems (external ladders/cradles etc.)

  - Medical/Clinical/Scientific
    - Correct operation of systems
    - Correct interfacing with adjacent systems (i.e. UPS back-up systems)
    - Connectivity to direct control centres

  - Escalators/lifts
    - Fire cause and effects
    - Correct operation of systems

- Correct interfacing with adjacent systems (i.e. fire alarms to fail safe)
- Connectivity to direct control centres
- Commercial ovens/kilns
  - Correct operation of systems
- Commercial fridges/freezers
  - Correct operation of systems

Guidance Note 57　Post-occupancy evaluation of buildings

Post-occupancy evaluation (POE) of a building not only demonstrates how well it is performing in use but also supports the design and construction of future new build and refurbishments by providing evidence-based knowledge, gained from past projects.

POE is becoming mandatory in many projects, and there are a variety of post-occupancy evaluation methods that can be tailored to individual needs.

There is a variety of best practice guidance available from RIBA, CIBSE, BSRIA, BRE, and more.

BS 8536-1[1] distinguishes between two types of aftercare period:

1. Initial aftercare (see BS 8536-1 clause 5.8.2.2)

2. Extended aftercare (See BS 8536-1 clause 5.8.2.3)

These aftercare periods are to enable understanding of facility performance, not to address defective works. During design and construction, measurement of the performance targets should take place through the process of formal information exchange. Information exchanges can occur within and at the end of project stages. POE is then the activity of measuring actual performance outcomes. If performance isn't as planned, remedial action can be taken alongside fine-tuning of systems and behaviours as required. Insight generated through POE also informs lessons learned. The project's Government Soft Landings (GSL) champion instigates, manages and evaluates each POE, although they may not carry out the POE activity themselves. The timing and format of the POE should be established as part of the GSL strategy and implementation plan.

In accordance with BS 8536-1, clause 5.8.2.3.4, formal POE of the building's performance should be conducted at the end of years one, two and three.

Organisations/departments may want to vary this to suit their own specific needs.

The POE should address those targets that have been set as performance outcomes (refer to Guidance sections 4.2–4.5 inclusive). POE methodologies should be appropriate and could include surveys, data collection enabled via

---

[1]　British Standards Institute (2015) "BS 8536-1:2015: Briefing for design and construction – Part 1: Code of practice for facilities management (Buildings infrastructure)" British Standards Institute.

---

building management systems (BMS) for example, and observations. In its simplest form, POE could consist of short review meetings with end-users, operators and relevant members of the project team. The POE should be recorded in a report considering all the performance targets determined for the project, with particular emphasis on:

- Operational management

- Durability and serviceability

- Performance of systems

- End-user experience

- Lessons learned

Note: Organisations/departments should decide whether POE activities are to be undertaken internally or carried out by an independent third party (in which case budgets will need to be established).

Guidance Note 58   Monitoring obsolescence

As assets are operated and maintained over time, it would be normal for a gap to emerge between the actual performance and value of the asset, and the performance and value necessary for the asset to have continued relevance. This gap is referred to as the obsolescence gap.

All assets will become progressively obsolete, and no longer fit for their originally intended purpose. The differential between fitness for purpose at outset and then at 5/10/50 years will obviously widen.

Scheduled maintenance work will reduce the speed of obsolescence, for example roof repairs to rectify wear and tear. Refurbishment work will also reduce the speed of obsolescence, for example the provision of new windows to rectify wear and tear and to enhance thermal efficiency.

Obsolescence gaps are often relevant in times of economic recession where investment in certain elements of the built environment is reduced, as other essential services are prioritised. It therefore, for example, becomes necessary to monitor obsolescence of required but unaffordable properties and develop a strategy for refurbishment and improvement as required.

Guidance Note **58**

Glossary

Alternative dispute resolution (ADR)	Ways of addressing contractual disputes without resorting to litigation. Mediation, conciliation, expert determination, adjudication and arbitration are all possible ADR methods.
Assess	The second project life-cycle stage where options and feasibility are assessed.
Assurance	The process of providing confidence to stakeholders that the project will meet the defined needs and benefits.
Behavioural procurement	An approach to procurement strategy, supplier selection, contract award and contract management that puts a high weighting on behavioural and cultural fit of the client and contractor team. A recurring theme in such approaches is the value attributed to openness and transparency. This starts with the client being open and transparent about the selection process and criteria and continues throughout the relationship with a focus on collaboration, joint problem-solving and the creation and sharing of knowledge.
Benefit	The measurable improvement resulting from an outcome, perceived as an advantage by one or more stakeholders, which contributes towards one or more organisational objectives. (Note: some outcomes from projects are perceived as a disadvantage by one or more stakeholders but the project is beneficial overall and is pursued. These are often referred to as dis-benefits.)
Built environment	Man-made assets and infrastructure, regardless of client type, funding, size, scale or complexity. Assets and infrastructure exist in transportation (road, rail, airports, maritime ports), power and utilities (nuclear, oil & gas, tidal lagoons, offshore wind, solar, water, electricity, telecommunications), natural defences (flood defences, dams) as well as buildings (homes, hospitals, schools, factories, warehouses, offices, hotels) and the parks, plazas and other spaces that create the environment in which people interact.
Business case	Provides justification for undertaking a project. It evaluates the benefit, cost and risk of alternative options and provides a rationale for the preferred solution. The business case is developed over time from an outline business case at the end of the Initiate stage to an intermediate business case at the end of the Assess stage to a full business case at the end of the Design stage.
Change control	A process that ensures potential changes to the deliverables of a project or the sequence of work in a project are recorded, evaluated, authorised and managed.

Code of Practice for Project Management for the Built Environment, Sixth Edition. Chartered Institute of Building.
© 2022 John Wiley & Sons Ltd. Published 2022 by John Wiley & Sons Ltd.

Client	Entity, individual or organisation commissioning and funding the project, directly or indirectly.
Client project manager	Responsible to the client sponsor for achieving project objectives. The project manager may be a client employee or consultant. In either case, the project manager ensures the administration of any contract(s) on behalf of the client.
Client sponsor	Accountable, on behalf of the wider client organisation, for achieving beneficial outcomes from the project including representing the needs of end users and funding bodies.
Collaboration theme	Addressing the need to build strong relationships across diverse networks of people and organisations, upholding inclusivity and equality of opportunities for work across the asset life cycle.
Consent	Permission to carry out a part of a project, for example planning permission.
Consenting strategy	The approach adopted on a project to gain the whole range of permissions necessary to undertake the project in compliance with the law.
Consultant	Specialist advisors to the client team, for example, architects, engineers, technology and process/method experts.
Contingency	A provision for a potential future situation. Contingency is typically financial or time-based, but can also refer to making provision for additional resources or spares.
Contractor	Responsible for delivering the design, build or maintenance of the physical asset, in whole, or in part, in line with the contract(s) administered by the client project manager.
Cost and budget planning and management	The work, once detailed planning of scope, quality, time and resource is complete, to create cost plans and budgets so that it is understood where capital and operational costs will be incurred over the life of the asset, and to track and manage expenditure over the life cycle.
Define	The third project life-cycle stage where the project approach and procurement strategy are defined.
Delegated limits of authority	Clarify the responsibilities of decision-makers in the project organisation. These responsibilities will relate to decisions about commitment of expenditure and legal undertakings, and about any other decision objectives defined.
Delivery model	The approach chosen that provides the best fit between the needs of the project and the availability and capability of resources in the supply chain. The term is derived from best practice guidance in sourcing public works projects.
Design	The fourth project life-cycle stage where the specifications and functionality for the asset are designed.
Design for Manufacturing and Assembly (DfMA)	Combines two methodologies, Design for Manufacturing (DfM) which focuses on efficient manufacturing, eliminating waste within a product design with Design for Assembly (DfA) which focuses on effective off-site or on-site assembly, minimising resources and disruption to other activities happening on site.
Design management	The ongoing processes, business decisions and strategies that enable innovation and create effectively-designed assets that meet defined needs and benefits.

Earned value analysis	A method of tracking progress by comparing the work achieved (earned) in the time and the work achieved (earned) for the money spent.
End user	Occupants or users of the built environment.
Environmental Impact Assessment (EIA)	The process of evaluating the likely environmental impacts of a proposed project and the assets it will create, taking into account inter-related socio-economic, cultural and human-health impacts, both beneficial and adverse.
Estimating funnel	A concept to describe how the accuracy of an estimate increases as more work is put into the definition of the project.
Golden thread	A phrase used to communicate the need for connected thinking and practices across the project life cycle to create value. Used specifically to refer to the need for a golden thread of building information.
Identify	The first project life-cycle stage where needs and benefits are identified.
Implement	The fifth project life-cycle stage where the design for the asset is used and constructed.
Information	Processed, organised data. According to the ISO19650 family of standards addressing information management in the built environment. Information can be structured (e.g. a drawing) or unstructured (e.g. a soil sample).
Information management	The practices to make sure the right information is delivered to the right destination at the right time to meet a specific purpose.
Integrated project insurance	A form of insurance that insures project risks rather than liabilities. It operates on a blame-free basis and insures outcomes rather than causes for the whole project organisation.
Knowledge theme	Addressing the need to create and control the use of knowledge, including intellectual property, to produce complete and accurate information to support the asset in use, over the whole life cycle and to ensure a culture of sharing, learning and continuous improvement.
Leadership theme	Addressing the need to focus on people, their skills, well-being and career opportunities, and also the need to provide competent governance and decision-making to lead the team though uncertainty and complexity.
Life cycle	The key stages of the whole life of a built asset and the objectives and decisions associated with each stage. The Code of Practice is structured around eight life-cycle stages that form a closed loop and address the work necessary to identify, assess, define, design, implement, validate, operate and retire assets.
Life-cycle management	The actions taken to manage and maintain an asset through the operations stage of the life cycle.
Milestone	A significant event/key point in the achievement of a project. Milestones may occur at the same time as the decision points between stages (stage gates) – see Stage below, or may be aligned with completion of key deliverables within a life-cycle stage. Milestones are not activities in a time plan/project schedule/project programme because they have no (zero) duration.
Obsolescence gap	The gap that inevitably emerges over time between the actual performance and value of the asset, and the performance and value necessary for the asset to have continued relevance.

Glossary

Output	A physical or knowledge-based asset that is produced, constructed or created as a result of a planned activity. Alternatively referred to as a deliverable.
Outcome	The result of change, normally affecting real-world behaviour and/or circumstances.
Operate	The seventh project life-cycle stage where the built asset is operated and maintained.
Operations and Maintenance Manual (O&M)	Provides all relevant information to the Operator of the asset based on the intended use and the client's strategy for life-cycle management.
Operator	Responsible for operation and maintenance of the asset as designed and built on behalf of the client and in compliance with all relevant legislation. *In some situations, the operator may be the same entity as the end user.*
Portfolio	A collection of projects or assets grouped together to facilitate their management to meet strategic objectives.
Productivity	Addressing the need to innovate and use appropriate modern methods of construction to optimise life-cycle cost and profitability, managing the pace of build and the achievement of quality and sustainability targets.
Project	In this *Code of Practice*, the term project describes the multiple ways over the entire life cycle of a built environment asset including the operational phase, in which clients organise the work to create, re-purpose and eventually retire built assets in order to achieve objectives and realise the desired value for end users, clients and funding bodies. Projects are delivered by temporary organisations: teams of people, from multiple organisations, working collaboratively in a structured way: ▪ to achieve defined objectives and quality standards, ▪ in a context of competing time and cost constraints, ▪ navigating significant uncertainty and risk, ▪ to operate in an environmentally friendly manner.
Project brief	A document produced in the Assess stage of the life cycle. It draws from the Project Mandate and is a key input to the Intermediate Business Case and the Project Execution Plan. The chosen concept for the project and defined outputs, outcomes and benefits are included in the Project Brief.
Project execution plan (PEP)	The collection of up-to-date plans and protocols for carrying out a project that is owned by the client project manager. In some environments, this is known as the project management plan (PMP), the project initiation document (PID) or the project handbook.
Project governance	The principles, policies and procedures by which a project is authorised and directed to accomplish agreed objectives
Project mandate	A document produced in the Initiate stage of the life cycle outlining the scope of the project expressed in terms of what it will deliver, and most importantly, what it will not deliver, by clearly defining its boundaries. A key input to the Business Case.

Project management approach	The choice of how to organise the delivery of project using a linear/ sequential approach, an iterative/agile approach or a hybrid.
Project organisation	The way in which project resources are structured to deliver objectives.
Project schedule	Time plan for a project or process. Note: on a construction project, this is often referred to as a 'project programme'.
Project stakeholders	The individual people or organisations/groups who are interested in, and can affect, or are affected by the project.
Qualitative risk analysis	A way of prioritising individual risks using qualitative variables such a likelihood, size of consequence, proximity or velocity.
Quantitative risk analysis	A way of determining the combined effects of all estimating uncertainty and risk events on the project's business case, whole-life cost, capital budget and/or timeline. This is known by different terminology in different companies, including:
	▪ Cost risk analysis (CRA)
	▪ Schedule risk analysis (SRA)
	▪ Cost and schedule risk analysis (CSRA)
	▪ Probabilistic risk analysis (PRA), or just
	▪ Quantitative risk analysis (QRA).
Quality planning and management	The work to define the standards to which the work will be completed and to protect this through change control to avoid requirements creep. Quality encompasses regulatory compliance and related targets and values related to health, safety and well-being of the project team, and of the operators and end users of the assets.
Quality theme	Addressing the need to clearly define standards and the associated requirements of clients and end users including, but not limited to, building safety, and to ensure and satisfy specifications correctly, the first time.
Resource levelling	Also known as resource-limited scheduling, is the process of making the most of the limited resources available. Resource levelling forces the amount of work scheduled not to exceed the limits of the resources available. This inevitably results in either activity durations being extended or entire activities being delayed until resource is available. Often this means a longer overall project duration. Using this method, cost is relatively more important than time.
Resource planning and management	The work to build a version of the time plan that can be resourced within the budget and to manage this over time. Related terms include resource smoothing and resource levelling: ways that the project manager makes sure the work to be done is delivered at the best possible time given the project objectives and resources available.
Resource smoothing	Also known as time-limited scheduling is the process of making sure resources are used as efficiently as possible and of increasing or decreasing resources as required to protect the end date of the project. Using this method, time is relatively more important than cost.

Retire	The eighth project life-cycle stage where the asset is re-purposed or demolished/destroyed.
Risk	The effect of uncertainty on objectives.
Risk appetite	The amount of and type of risk that an organisation is willing to take in pursuit of objectives.
Risk capacity	The amount and type of risk that an organisation is able to take in pursuit of objectives.
Risk register	A repository for storing information about risks.
Risk theme	Addressing the need to explicitly deal with uncertainty and ambiguity to improve predictability and the ability to meet objectives.
Scope planning and management	The work to define what work will be done and to ensure this is protected during implementation using change control to avoid scope creep.
Soft landings	A methodology to ensure operational needs are fully considered and appreciated at the design stage and embedded in procurement and contractual obligations.
Stage	In this Code of Practice, the term stage is used to describe each part of the project life cycle. This may be used interchangeably with phase.
Stage gate	The decision points between stages are commonly referred to as stage gates. An alternate term is decision gate.
Sustainability theme	Addressing the need to be stewards of the natural world, addressing air quality, climate change, water usage, land usage, resource usage and biodiversity and creating an effective circular economy, to step away from a paradigm of 'make, use, dispose'.
System	A set of things working together as part of an integrated whole.
Sub-system	A self-contained system that is required to be integrated with a wider system.
Time planning and management	The work to define a realistic time plan for the project, defining the work that is most critical in driving completion (the critical path) and working to protect delivery performance. Time plans are commonly referred to as the project schedule or project programme.
Validate	The sixth project life-cycle stage where the implemented design is integrated and handed over to operations.
Value theme	Addressing the socio-economic value of the asset in terms of whole-life costs and benefits and ensuring there is a competent process to decide on the relative priorities of drivers of value including economic and social objectives, to procure and measure performance to validate priorities.

Past working groups of *Code of Practice for Project Management*

Fifth Working Group for the Revision of the *Code of Practice for Project Management*

Saleem Akram BEng (Civil) MSc (CM) PE FIE NAPM FIoD EurBE FCIOB — Director, Construction, Innovation and Development, CIOB

Colin Bearne — Gardiner & Theobold

Sarah Beck MRICS NAPM — Royal Institute of British Architects

Andrew Boyle — Tesco

Shaun Darley — Voice of Reason Ltd/MB PLC

John Eynon — Open Water Consulting

Dr Chung-Chin Kao — Innovation & Research Manager, CIOB

Scholarships & Faculties Manager, CIOB

Una Mair — CIOB Trustee

Institution of Civil Engineers

Gavin Maxwell-Hart BSc CEng FICE FIHT MCIArb FCIOB — ARUP

Technical Editor

Alan Midgely

Arnab Mukherjee BEng(Hons) MSc (CM) MBA MAPM FCIOB — Turner & Townsend

Development Manager, CIOB

Paul Nash MSc FCIOB — University of the West of Scotland

Piotr Nowak MSc Eng. — Salford University

Dr Milan Radosavljevic UDIG MIZS-CEng ICIOB — Chair Working Group

Eric Stokes MCIOB FHEA MRIN — University College London

David Woolven MSc FCIOB — College of Estate Management,

Royal Institute of Chartered Surveyors,

Roger Waterhouse MSc FRICS FCIOB FAPM — Association of Project Management

The following also contributed in development of the fifth edition of the *Code of Practice for Project Management*

Andrew Barr — Davis Langdon

Richard Biggs MSc FCIOB MAPM MCMI — Construction Industry Council

Northumbria University at Newcastle

Richard Humphrey FCIOB FRSA FCMI FIoD MAPM PGCert FHEA EurBE — Chair, Health & Safety Advisory Committee

Vaughan Burnard — Head of School of the Built Environment & Engineering Faculty of Arts, Environment & Technology, Leeds Metropolitan University

Professor Farzad Khosrowshahi FCIOB — URS

Consultant, CIOB

University College London

Publisher, Wiley-Blackwell, John Wiley & Sons Ltd, Oxford

Dean Hyndman

Dr Sarah Pearce BA (Hons) MSc

Dr Aeli Roberts MSc GDL BVC ICIOB

Dr Paul Sayer

Code of Practice for Project Management for the Built Environment, Sixth Edition. Chartered Institute of Building.
© 2022 John Wiley & Sons Ltd. Published 2022 by John Wiley & Sons Ltd.

Fourth Working Group for the Revision of the *Code of Practice for Project Management*

Saleem Akram BSc Eng (Civil) MSc (CM) PE MASCE MAPM FIE FCIOB	Director, Construction, Innovation and Development, CIOB
Alan Crane CBE CEng FICE FCIOB FCMI	Chair, Working Group, Vice President CIOB
Roger Waterhouse MSc FRICS FCIOB FAPM	Vice Chair; Working Group, Royal Institution of Chartered Surveyors Association for Project Management
Neil Powling DipBE FRICS DipProjMan(RICS)	Royal Institution of Chartered Surveyors
Gavin Maxwell-Hart BSc CEng FICE FIHT MCIArb FCIOB	Institution of Civil Engineers
John Campbell BSc (Hons), ARCH DIP AA RIBA	Royal Institute of British Architects
Martyn Best BA Dip Arch RIBA MAPM	Royal Institute of British Architects
Richard Biggs MSc FCIOB MAPM MCMI	Construction Industry Council
Paul Nash MSc MCIOB	Trustee of CIOB
Ian Caldwell BSc BARch RIBA ARIAS MCMI MIOD	
Professor James Somerville FCIOB MRICS MAPM MCMI PhD	
Professor John Bennett DSc FRICS	
David Woolven MSc FCIOB	
Artin Hovsepian BSc (Hons) MCIOB MASI	
Eric Stokes MCIOB FHEA MRIN	
Dr Milan Radosavljevic UDIG PhD MIZS-CEng ICIOB	
Arnab Mukherjee BEng(Hons) MSc (CM) MBA MAPM MCIOB	Technical editor

The following also contributed in development of the fourth edition of the *Code of Practice for Project Management*

Keith Pickavance	Past President, CIOB
Howard Prosser CMIOSH MCIOB	Chair, Health & Safety Group, CIOB
Sarah Peace BA (Hons) MSc PhD	Research Manager, CIOB
Mark Russell BSc (Hons) MCIOB	Co-ordinator, Time Management Group, CIOB
Andrzej Minasowicz DSc PhD Eng	Vice Director of Construction Affairs, Institute of Construction Engineering and Management, Civil
FCIOB PSMB SIDiR	Engineering Faculty, Warsaw University of Technology
John Douglas FIDM FRSA	Englemere Ltd
Dr Paul Sayer	Publisher, Wiley-Blackwell, John Wiley & Sons Ltd, Oxford

Third Working Group for the Revision of the *Code of Practice for Project Management*

F A Hammond MSc Tech CEng MICE FCIOB MASCE FCMI	Chairman
Martyn Best BA Dip Arch RIBA	Royal Institute of British Architects
Alan Howlett CEng FIStructE MICE MIHT	Institution of Structural Engineers

Gavin Maxwell-Hart BSc CEng FICE FIHT MCIArb	Institution of Civil Engineers
Roger Waterhouse MSc FCIOB FRICS MSIB FAPM	Royal Institution of Chartered Surveyors and Association for Project Management
Richard Biggs MSc FCIOB MAPM MCMI	Association for Project Management
John Campbell	Royal Institute of British Architects
Mary Mitchell	Confederation of Construction Clients
Jonathan David BSc MSLL	Chartered Institution of Building Services Engineers
Neil Powling FRICS DipProjMan (RICS)	Royal Institution of Chartered Surveyors
Brian Teale CEng MICBSE DMS	
David Trench CBE FAPM FCMI	
Professor John Bennett FRICS DSc	
Peter Taylor FRICS	
Barry Jones FCIOB	
Professor Graham Winch PhD MCIOB MAPM	
Ian Guest BEng	
Ian Caldwell BSc Barch RIBA ARIAS MIMgt	
J C B Goring MSc BSc (Hons) MCIOB MAPM	
Artin Hovsepian BSc (Hons) MCIOB MASI	
Alan Beasley	
David Turner	
Colin Acus	
Chris Williams DipLaw DipSury FCIOB MRICS FASI	
Saleem Akram B Eng MSc (CM) PE FIE MASCE MAPM MACost E	Secretary and Technical Editor of third edition
Arnab Mukherjee B Eng MSc (CM)	Assistant Technical Editor
John Douglas	Englemere Ltd
David Woolven MSc FCIOB	

First and Second Working Groups of the *Code of Practice for Project Management*

F A Hammond MCs Tech CEng MICE FCIOB MASCE FIMgt	Chairman
G S Ayres FRICS FCIArb FFB	Royal Institution of Chartered Surveyors
R J Cecil DipArch RIBA FRSA	Royal Institute of British Architects
D K Doran BSc Eng DIC FCGI CEng FICE FIStructE	Institution of Structural Engineers
R Elliott CEng MICE	Institution of Civil Engineers
D S Gillingham CEng FCIBSE	Chartered Institution of Building Services Engineers
R J Biggs MSc FCIOB MIMgt MAPM	Technical editor of second edition and Association for Project Management
J C B Goring BSc (Hons) MCIOB MAPM	
D P Horne FCIOB FFB FIMgt	
P K Smith FCIOB MAPM	
R A Waterhouse MSc FCIOB MIMgt MSIB MAPM	
S R Witkowski MSc (Eng)	Technical editor of first edition
P B Cullimore FCIOB ARICS MASI MIMgt	Secretary

For the second edition of the *Code* changes were made to the working group which included:

L J D Arnold FCIOB

P Lord AA Dipl (Hons) RIBA PPCSD FIMgt (replacing R J Cecil, deceased)

Royal Institute of British Architects

N P Powling Dip BE FRICS Dip Proj Man (RICS)

Royal Institution of Chartered Surveyors

P L Watkins MCIOB MAPM

Association for Project Managers

Index

Index